29286

MI

This bc

23 AP

17 MA

27 MAY

22 FEB

JUN 20

23 FEB 20

-7 I

Sustainable Development

Sustainable Development
Constraints and Opportunities

Mostafa Kamal Tolba

Butterworths
London Boston Durban Singapore Sydney Toronto Wellington

British Library Cataloguing in Publication Data

Tolba, Mostafa Kamal
  Sustainable development : constraints
  and opportunities.
  1. Economic development   2. Human ecology
  I. Title
  330.9   HD82

  ISBN 0-408-00877-6

Library of Congress Cataloging in Publication Data

Tolba, Mostafa Kamal.
  Sustainable development.

  Includes index.
  1. Environmental protection.   2. Environmental
policy.   3. United Nations Environment Programme.
I. Title.
TD170.3.T66   1987   333.7'2   87-5173
ISBN 0-408-00877-6

First published in 1987 by Butterworth Scientific,
PO Box 63, Westbury House, Bury Street,
Guildford, Surrey GU2 5BH, England

Telephone: Guildford (0483) 31261   Telex: 859556 SCITEC G

Photoset by Prima Graphics, Camberley, Surrey
Printed and bound in Great Britain by
Biddles Ltd, Guildford and King's Lynn

Contents

Preface

Our understanding of the interrelationships between environment and development has undergone a profound change during the past 15 years. At the end of the 1960s, it was generally believed that it was possible to have either one or the other, but not both simultaneously. In other words, if we wanted development, the price to pay would be a loss in terms of environmental quality, and *vice versa*.

That view has now been completely overturned with the realization that environment and development are interdependent, and are in fact mutually supportive. It is now clear that without environmental protection, it is not possible to have sustained development, and without development, it is not possible to sustain a high quality of our environment and an improved quality of life for all the world's citizens. Thus, what we need is sustainable development, that is development that can be sustainable over the long term by explicitly considering the various environmental factors on which the very process of development is based.

This book contains a selection of my statements made between 1982 and 1986. As may be expected, during this 5-year period our knowledge of environment–development relationships has grown considerably, and this evolving perception can be discerned in my statements spanning those years. Taken together, the main thread binding all my statements presented in this volume is the fact that long-term development can only be achieved through sound environmental management, that is, sustainable development, which is the title of this book.

A final word of apology. These speeches were delivered in my capacity as the Executive Director of the United Nations Environment Programme. They thus cover primarily broad areas of interest of UNEP's programmes and policies. For this reason the reader will notice some duplication of the issues dealt with. To a large extent, this problem has been kept to an absolute minimum through careful editing. However, in some cases it was not possible to do so without seriously interrupting the flow. I hope the reader will bear with me on this.

Nairobi, February 1987   *Mostafa Kamal Tolba*

Acknowledgements

Ideas and concepts often develop on the basis of discussions with colleagues. Many of the ideas and concepts outlined in this volume owe much to my colleagues, both past and present, and many scientists and policy-makers from all over the world. I am truly grateful for their past and continuing generous assistance and cooperation.

I would especially like to express my deep gratitude and thanks to my colleague and friend, Dr Asit K. Biswas, President of the International Society for Ecological Modelling, Oxford, England, who has been one of my Senior Scientific Advisors since the beginning of UNEP, for helping me to select and edit the speeches in this book.

Environmental protection

Address to the Japan Environment Agency Symposium

Tokyo, Japan, January 1982

Japan's problems centre chiefly around air and water pollution. These are Japan's special concerns – which she shares with the world's other highly developed nations. However, industrial contamination is only one aspect of environmental deterioration; other countries are facing formidable problems of a different kind. These include spreading deserts, tropical deforestation, soil erosion, cities growing out of control, disease-ridden water, and energy and food shortages. These international environmental problems also affect Japan.

In the developed nations environmental destruction is largely the consequence of prosperity. The position is reversed in the poor, developing nations where environmental impoverishment is caused mainly by poverty itself and, unlike the developed nations, they lack the means to tackle their problems.

Poverty is locking the people of the Third World into a dismal cycle of events: in their efforts merely to meet basic needs of food, shelter and heat, they are being forced to destroy the very resources on which their future survival (and the future prosperity of all) depend.

In the face of the current global recession, the temptation is for the industrialized nations not to regard the environmental destruction taking place in faraway lands as an immediate problem and, instead, to relegate it to a position of secondary importance. This is dangerous. A country like Japan, which must import almost 75% of her timber from abroad, virtually all her minerals, 99% of her oil and other so-called 'fossil' fuels and a considerable quantity of her foodstuffs, is directly affected by the wastage of resources taking place in these far-off places.

If, to meet the demands of Japan's hardwood industry, a stretch of tropical forest is felled, and the soil subsequently washed away, then whose concern is that – Japan's of that of the tropical country involved? The answer is both.

This is a straightforward example of identity of interest. Less obvious perhaps is the fact that, as resources become exhausted, so national economies suffer; and as nations slide deeper into debt so, inevitably, up go the trade barriers. Today, manufacturing nations face the dual threat of a dwindling supply of raw materials and closing markets.

The aims of the environment movement are the same as those of the North–South dialogue. The idea behind that dialogue is that it pays the rich nations to invest in the poor. These are not the politics of the begging bowl, but of international justice and benign self interest.

The time has now come for Japan to apply the same commitment and resolve she has shown in tackling her problems at home to the international scene. An opportunity is coming for Japan to make that commitment. In Nairobi in May 1982 a conference will be held on the global environment, open to all member states of the United Nations. To be called the 'Session of a Special Character', the conference is a sign of the concern of the United Nations Environment Programme, endorsed by the UN General Assembly, that the international community is not doing nearly enough to solve common environmental problems. The occasion for the 'Session of a Special Character' is the tenth anniversary of the UN Conference on the Human Environment in Stockholm, at which 113 nations declared their intention to 'safeguard and enhance the environment for present and future generations of man'. The Nairobi conference will provide Heads of State, top government ministers and representatives of leading international organizations attending the conference with a once-in-a-decade chance to put the environment at the top of the international agenda for action.

The Stockholm meeting approved an Action Plan of over 100 separate recommendations and created UNEP to catalyse nations into implementing those recommendations. UNEP has produced a series of reports for the Nairobi meeting which review the international performance in meeting the goals of the Stockholm plan. The issues covered in the UNEP reports coincide with my brief to review the changes in the state of the environment, bringing us up to the present day, and to discuss future trends and suggested solutions for the problems we foresee.

It is not my intention to present an unduly bleak picture. There have been some encouraging developments in the intervening 10 years. One of the most positive has been a clear awakening among governments as to the international dimension of environmental problems. For example, a series of regional and international environment agreements have been concluded and I am pleased to note that Japan has become a member of most of them. But this can only be a start; much more needs to be done.

At the end of last year a group of leading Japanese decision makers, chaired by former foreign affairs minister Dr Saburo Okita, met to suggest the course for Japan's future environmental policies. The group called on the Japanese government to step up its efforts and to prevent a further deterioration of the global environment. An interim report issued by the group stated that 'As global environmental policies have an impact on the great majority of the world's people, it is ultimately in the interest of all nations to preserve the environment'. UNEP wholeheartedly endorses this preliminary conclusion of the Committee, and at the Nairobi conference we will be looking to Japan to put the call into effect.

An area of special concern to Japan is conservation of marine resources. A UNEP review of the changes in the state of the environment over the previous 10 years notes that while there is no evidence of serious pollution of the open oceans, coastal waters have been exposed to increasing threats from industrial contamination and unsound development. With fish playing a major role in the Japanese diet, this should be a source of great concern since 90% of the world's marine catch comes from coastal waters.

UNEP has identified the areas most at risk. We call these 'hot spots'. They tend to be estuaries, heavily industrialized bays and semi-enclosed seas such as the Mediterranean or Seto Inland Seas. Scandinavia's Baltic Sea is one of the areas most at risk: throughout the decade the area of Baltic seabed devoid of all life has continued to spread at an alarming rate. We can have deserts at sea, as well as on land.

Japan continues to experience similar problems despite the introduction of strict anti-pollution laws. Red tides still persist, and today Tokyo and Ise Bays rank among the world's most seriously polluted coastal waters.

The continuing pollution of Japanese territorial waters is a good illustration that national action, no matter how strict, no matter how well enforced, can never be good enough: crude oil accounts for 75% of the pollution incidents in Japanese waters, nearly all of this coming in from the open seas on the Kuroshio and other dominant currents. UNEP calculates that between 1970 and 1978 well over a million tonnes of crude oil were spilled on the world's oceans. With many countries now exploiting offshore oil reserves this figure is likely to rise over the next decade.

What is the answer? On marine resources – and indeed on every aspect of the management of natural resources – we must make environmentally sound decisions that enable us to continue to exploit those resources, indefinitely in the case of the so-called 'renewable' resources such as fish or forests and for the maximum time possible in the case of exhaustible or 'finite' resources such as metal ores or oil.

This is what is now called sustainable development – a strategy of economic common sense. From the beginning, UNEP's call has been 'development without destruction', and every tactical solution I shall outline in this address is rooted in this overall strategy of continuous or sustainable development.

An important priority for nations over the next 10 years will be to plough more effort into research in every aspect of the environment. Without research we cannot make environmentally sound decisions. For example, most fishing in tropical waters is carried out in ignorance of how great a catch these warm waters can sustain.

Mostly for reasons of sheer bad management the world's fish catch is now about 10% less than it might otherwise have been. One reliable estimate is that, with good management, the world's fish catch could be doubled, even trebled. To prevent a further decline, Japan and the other fishing nations must observe scientifically sound fishing quotas. Fish and other living resources can be surprisingly resilient – a ban on Europe's North Sea herring allowed stocks to recover with the result that a limited harvest is now once again possible. And to lighten the pressure on the marine catch, UNEP is advocating development of more mariculture and aquaculture. Our report notes that China's fish farm harvest was 450 000 tonnes in 1978. Other nations would do well to follow suit.

Pollution control and fishery management are two important components of UNEP's Regional Seas Programme: for the Caribbean, Arab Persian Gulf region, the Asean area, the Mediterranean and the Atlantic coastal waters of Central and West Africa, UNEP has succeeded in getting nations – some of which have wide political differences – to agree on common environment Plans of Action. Other accords are now being negotiated. Since the fishing fleets of Japan, and those of other major fishing nations, ply many of these waters it would serve their national interests to support these environment agreements.

The third major component of UNEP's Regional Seas Programme is protection of fish and crustacea breeding and feeding areas. Strictly observed quotas and pollution laws will mean little unless these areas – which include marshes, estuaries, mangrove swamps, coral reefs and so on – are spared destructive coastal development. Here UNEP is advocating formulation of guidelines for coastal development, including offshore mineral extraction and the creation of marine conservation areas. Other nations should follow Japan's example and join the poorly supported international Wetlands Convention. Not only as fish breeding areas, but also as a control of inland drainage, a habitat for waterfowl and so on, these wetland areas are as much a part of a nation's assets as, say, an oil field or an industrial complex.

Like the marine environment, the atmosphere is a topic of global concern. And again I return to my central theme: while the quantity of polluting substances Japan puts into the atmosphere has fallen, elsewhere in the world – for the most part – the trend has been the reverse. We know there has been a steady build up of carbon dioxide in the atmosphere, resulting from the burning of more oil, coal, shale oil and from the destruction of forests, but we do not know what effect it will have. It could cause what has been termed the 'greenhouse effect', which might result in a global warming. Such a rise in temperature would certainly alter the climate and affect food production. Most energy scenarios predict that the world will burn an even greater quantity of fossil fuels – chiefly coal – over the next 10 years, so the likelihood is that the amount of carbon dioxide in the atmosphere will continue to increase, and at a faster rate than over the previous decade.

Fossil fuel burning also releases sulphur and nitrogen oxides into the air. Though prevailing winds may blow these pollutants many hundreds of kilometres away from their source, they eventually fall to earth, diluted in rain. This is acid rain, which is killing freshwater lakes in Scandinavia and North America. The acid content of some Swedish lakes has increased 100-fold over the past 30 years, and this is just one measure of the kind of transboundary pollution problems we are facing today.

Another measure is the perceived threat to the ozone layer which protects us from carcinogenic ultra violet light. The threat to the ozone layer comes from substances called chlorofluorocarbons emitted by spray cans, refrigerators and other devices. Within the limits of present detection measures, we cannot say with any certainty whether a depletion of the ozone layer is occurring. The urgent need now is to turn these uncertainties and theories into hard facts and figures. This can only be achieved through research and monitoring. But the threat is real enough for some nations to have demanded the creation of an ozone treaty.

The well-being of our atmosphere depends not only on the effectiveness of pollution control measures, but also on energy policies. During the previous decade world energy consumption rose by a third, and demand will continue to increase. UNEP advocates that countries should develop energy policies based upon the most appropriate mixes of energy types to meet their increasing demands while minimizing the threat to the environment. Industrialized nations are urged to make savings in energy use through conservation. Energy efficiency will not hold up economic growth. For example, in FR Germany between 1973 and 1980 the gross national product grew 20%, while energy consumption over the same period rose by less than 3%.

UNEP also hopes to see the rapid development of appropriate and environmentally sound renewable technologies, such as solar, hydro and biogas power. To meet firewood demands UNEP is asking for a target to be set of a four-fold increase in fuelwood plantations, to be established on wastelands and other areas of little agricultural value.

Over the next 10 years the nuclear debate will continue to be a matter of controversy. Today, nuclear power is only meeting 6% of world energy needs, with some 90% of nuclear power being generated in the Western industrialized nations and Japan. Despite strong antinuclear movements in these countries, the trend will be for nations to continue with nuclear power development, at least in the short term. The emphasis for future policies should be on devising safe methods for disposal of nuclear waste.

Freshwater represents about 0.1% of all the water on the planet. The 1970s saw a staggering increase in demand on those limited freshwater supplies for industrial, agricultural and domestic uses. In Japan demand, which stood at 50 million cubic metres per day in the mid-1960s, increased to 120 million in 1973. But demand levelled out by the end of the 1970s as Japan began to recycle more and more waste water. However, this came too late to prevent 2% of the surface area from being affected by ground subsidence.

Japan could win many friends by sharing her recycling expertise with foreign powers. As would be expected, water supply problems are very severe in dry regions; in some parts of West Asia demand is now so great that fossil groundwater supplies are being withdrawn.

River basins, underground aquifers and so on are often shared by different nations. Pollution or exhaustion of these common supplies could become a source of regional antagonism. To cope with such problems, UNEP is promoting the development of internationally agreed guidelines for water management. There is also a need for regular assessment of global water needs.

Nineteen eighty-two is the second year of the International Drinking Water Supply and Sanitation decade. A large proportion of the diseases affecting the people of the developing world are waterborne, and only 29% of the people living in these countries currently have access to safe water supplies. The target of 'clean water for all' by 1990 looks difficult to achieve, but this should not offset a determined implementation of proposals to obtain this goal.

One serious environmental health problem to have emerged during the last decade is insect resistance to pesticides. For example, the malaria-transmitting mosquito has been making a sinister comeback. In Northern countries, most environmental diseases relate to an unacceptable level of pollutants in working and living surroundings. Japan's record in combatting environmental health hazards is exemplary in this regard. But, again, tough controls at home are not the only answer: evidence has emerged recently that some multinationals have been avoiding strict domestic regulations by exporting hazardous wastes to developing nations where controls are lax or non-existent.

The lack of international regulations has begun to rebound. According to a US government agency, half of all the coffee beans imported into the USA are contaminated by pesticides banned at home. To deal with such problems international chemical safety programmes and stricter controls on the disposal of wastes are required.

UNEP is concerned that, under the present economic recession, attention turned to crisis problems of runaway inflation, trade deficits, slowing rates of economic growth and mounting unemployment will be at the expense of programmes for environmental protection. In the rush to industrialize, developing nations have been paying less attention than required to environmental protection measures. These trends are likely to continue, unless the international community takes immediate and forceful action.

UNEP has produced a potent new argument in favour of environmental protection. A programme of cost–benefit evaluation has shown that environmental protection can be shown to pay in hard cash terms. For example, the results of a French study on 24 pollutants revealed that the cost of pollution in 1978 came to about 4% of the gross national product (GNP), whereas cost–benefit analysis revealed that an investment in environmental protection of up to 2% of GNP could have saved those additional costs. Similar studies have produced parallel results in other countries.

Another study – this time carried out by the International Labour Organization (ILO) – has shown that the stringent pollution control measures introduced by

Japan during the 1970s have encouraged her industry to renovate older polluting and wasteful plants. As a result, the study points out, Japanese industry is now cleaner and more efficient than that of many of her industrial competitors.

Deforestation, desertification and associated soil erosion problems are further areas of global concern. Despite slowing rates of tropical forest destruction, UNEP estimates that between now and the end of the century about 12% of the existing tropical forest cover – that is 114 million hectares – will go. In fact, the problem of forest destruction may be more serious than statistics based on satellite images reveal: in savannah areas, villagers using simple hand axes have already destroyed much of the secondary vegetation which, in the normal course of events, would have replaced the mature trees when they die. A survey carried out in Kenya has revealed that, at current rates of destruction, the country will have exhausted all its firewood resources within 10 years.

Once the tree cover is removed, the soil – which may have taken nature a milennium to form to a spade's depth – is often swiftly washed or blown away. That soil frequently ends up as silt in thousands of muddy rivers, carrying their unwanted cargo to the sea. That silt can, in turn, block the water flow and cause disastrous flooding.

No one has attempted to put a price tag on the cost to the international community of the soil lost each year – it could be two or three times Japan's gross national product. As part of its international strategy, UNEP is promoting the formulation of a global soils action plan.

Five years ago the international community agreed another Action Plan, to push back the advancing deserts. But governments have done little to implement the plan and deserts continue to exact an awful toll: currently 6 million hectares of the world's arable land is converted every year into desert, and a further 20 million hectares is degraded. Japan is uniquely well placed to take a world lead in putting the desertification Action Plan to work.

UNEP's solutions to these land-based problems are set out in numerous documents that have been distributed to governments. They find their most cogent expression in the World Conservation Strategy, which sets out UNEP's overall plan for continuous exploitation of ecosystems.

When talking about stopping the deserts, preserving marine resources and so on, we are also talking about human survival. The world's population is more than four billion and there are certain to be almost half as many people again to cater for by the turn of the century. For many of the world's nations the task is not to increase prosperity but to provide basic needs, such as health care, shelter, fuel, and the most fundamental need of all – food.

According to the United Nations Food and Agriculture Organization (FAO), virtually all the world's prime agricultural land has already been used up. FAO estimates that a population four times greater than Japan's is chronically underfed. A 60% increase in food production is needed between now and the end of the century. So crop production methods which make better use of the land already being farmed must be devised, post-harvest losses which could boost food production by a third in some Asian countries must be stopped, and wild habitats, which are the source of the genetic material that helps scientists keep one step ahead of pests and diseases, must be preserved.

Nations must also improve land-use planning to prevent agricultural land being squandered. Each year about 5000 square kilometres are lost to the tarmac and concrete of our expanding cities.

One of the most exacting challenges facing the international community over the next 10 years will be to meet the needs of human settlements which are exceeding their environmental limits. Over the past 10 years the urban population has increased by 450 million and today stands at 1800 million. This trend will continue.

Nearly all the growth has taken place in the cities of the developing world – today there are 22 Third World cities with a population of 4 million or more, and by the year 2000 there will be three times that number. One third of that urban population lives in slums or shanty towns, and many of the cities are experiencing a carbon copy of the pollution problems which, since the times of the Industrial Revolution, have degraded the quality of life in the cities of the developed world.

An important opportunity presents itself for Japan to share her experience in tackling the problems of urban environmental deterioration. If we are to have any hope of improving the quality of life in the cities, nations will have to give much more emphasis to the development and application of environmentally sound systems of human settlement, planning, building technology and service provision.

This address presents a generalized account of what UNEP sees as the main environmental problems the international community will face over the next 10 years. UNEP's solutions all come down to the common denominator of sustainable development.

One piece of the environment jigsaw is missing: this is peace and security. Though numerous arms limitation agreements testify to a widespread desire to stop the arms race, it continues apace. Currently the world is spending $500 billion each year on weapons. Most of that money goes into making weapons ever more efficient, ever more deadly. The destruction of Vietnam's mangrove forests by chemical warfare has given us a glimpe of the kind of devastation modern weapons can cause. Those chemicals turned the mangroves into a wasteland, and experts tell us a nuclear bomb would probably have done less lasting damage.

Japan is the only country to have experienced a nuclear attack, and to know the wasteland an atom bomb can cause. The process of environmental destruction can cause, and is causing, damage of a similar magnitude – slower perhaps, but no less terrible in its consequences.

In May 1982, at the Session of a Special Character, Japan and the other nations of the world will be presented with a unique opportunity to renew their commitment to tackling pollution, desertification, deforestation and the other destructive forces that are relentlessly pushing forward the frontiers of the wasteland. If Japan were to apply the same commitment she has shown in dealing with her domestic problems to the international scene then she could play a major role in making the world at the turn of the century a safer and more prosperous one than we find today.

2

Environment and development

Statement to the
Japan Advisory Committee to Study
Global Environmental Programmes

Tokyo, Japan, January 1982

Our understanding of the relationship between environment and development has undergone a profound change over the past 10–15 years. At the end of the 1960s, it appeared to all but the most enlightened that you either had one or the other, and that if you wanted to have development then the price to pay would be a loss in environmental quality.

That view has been overturned as we have come to realize that the two – environment and development – are interdependent. Without conservation you cannot have development, and without development you cannot have conservation.

International approval for this principle was given by the 113 nations attending the 1972 Stockholm UN Conference on the Environment. Since then terms such as 'ecodevelopment', 'sustainable development' and 'environmentally sound development' have come into widespread use. But how much are these terms really understood? How great is the understanding of the complex relationship between environment and economic development?

Long-term development can only be achieved through sound environmental management. Unfortunately, 10 years after Stockholm, the mass of the general public and a sizeable number of those who make development decisions are not aware, or at best dimly aware of this. Environmentalists have a duty to explain, and keep on explaining, the advantages of environmentally sound development.

UNEP's consistent plea since its establishment in 1973 has been 'development without destruction', and it has tried to justify this on the basis that a strategy of a sustained, rational use of natural resources is cost-effective and economic common sense.

Natural resources are of two kinds: renewable or depletable. Renewable resources are those that grow like fish and forests or which are replaced by the undisturbed workings of the natural world such as soil or oxygen. As long as the sun continues to shine and to provide us with energy (current calculations put this at around 4000 million years) these renewable resources need never be used up.

Depletable resources are oil, coal, mineral ores and so on. In most cases these resources are being used up much faster than they need be. Used in a productive, non-wasteful manner most of these resources could last for centuries to come. Despite a growing awareness of the need to conserve these resources they continue to be squandered. The problems posed today by a dwindling supply of natural resources are serious; by the turn of the century, now less than 20 years away, the problems we face today will seem insignificant by comparison.

Japan, which in 1979 spent a sum approaching ¥600 billion on various forms of development assistance, could help the developing countries restore their renewable resource base by making the environment a first consideration in her overseas aid policy. The major international development assistance agencies have made such a commitment. Countries such as Sweden and FR Germany have already taken a similar step. Japan should follow this lead.

A world-wide application of environmentally sound development will avert the developing countries' crisis. We know the problems, we also know the solutions. But these nations, led by the example of the industrialized countries, must show that they have the political commitment, that is the political will, to solve those problems.

The environment cannot be categorized, it cannot be placed in its own pigeon-hole. The environment has relevance to every aspect of man's political, social and economic activities. It cannot be relegated to a position of secondary importance while nations attempt to solve pressing problems of economic recession.

Against the march of time, and within the limits of inadequate resources, UNEP and its responsible partners in the environment movement are attempting to persuade nations of the short- and long-term benefits to be gained from sound environmental management of natural resources. When this is realized, then, simultaneously, the political commitment sought will be generated. We have a world to win. Much that has already been lost can be regained.

If nations show political commitment we will be able to stop the advancing deserts and 'regreen' the areas already lost; we can replant the trees and use the forests more wisely; we can combat marine pollution and restock depleted fisheries; we can stop soil erosion and make better use of the land already being farmed; we can prevent air pollution and improve the quality of life in exploding cities. Above all, we will be able to tackle poverty – the cause of most of our environmental problems.

The politicians and other decision makers we have to convince are not for the most part people with a background in the physical sciences. They tend to be economists, lawyers, social scientists, bankers, businessmen, planners and the like. Let us be realistic: these people do not make decisions with the long-term interests of the next generation in mind. They are made with a view to the next election, or to the annual balance or the next meeting of shareholders.

A distinguished British prime minister, bored by a lengthy exposition by a fellow politician on the long-term advantages of a particular strategy, finally lost patience and said: 'In the *long term* gentlemen, we'll all be dead'.

An investment in environmental protection pays – now. That is the message we have to get over to these decision makers and to the general public at large. UNEP has arguments, hard facts and figures with which to promote that message.

Certain concepts have emerged during the last decade in our understanding of the environment. There is increasing evidence that sustainable development is not realizeable unless emphasis is placed at a very early stage on a number of relationships. The first is the relationship between the natural world and human society and its development. The component parts of the natural world – forests, atmosphere, soil, the sea and so on – together make up the system which supports all life on earth. It is the strange habit of mankind always to abuse this life support system on which the survival and future prosperity of all depend.

A series of world conferences on food, population, human settlements, water, desertification and environmental education, held after the 1972 Stockholm conference, have served to broaden our realization that everything is related to everything else. The global environment makes all activities interdependent. For example, the problems of atmospheric ozone depletion, possible climatic warming due to increasing levels of carbon dioxide, acid rains and nitrogen shortages in the soil seem very different in character, but are in fact linked closely through the global cycles of carbon, oxygen, nitrogen and sulphur.

The focus of concern has come increasingly to rest on the effect of human activities on the environment and on the availability of natural resources and energy. Environmental management is concerned not so much with the improvement of the environment as with the management of all man's activities which depend on the resources of the environment, and which have an impact, beneficial and detrimental, upon our surroundings. Its aim is to ensure an approach in which economic development and conservation of natural resources are pursued as goals of equal importance.

What we need now are new and imaginative development policies. At the national level one could envisage measures to restrict excessive or wasteful consumption of resources through alternative lifestyles and development patterns. At the international level it is necessary to anticipate environmental threats and identify measures to control them. Simultaneously, management of common resources, such as the atmosphere or the open oceans, should be geared to urgent international development needs and priorities.

The second relationship which must be emphasized if sustainable development is to be achieved is the connection between economic growth and development. Recent years have witnessed a broadening of our understanding of this relationship. Development is no longer seen exclusively as a matter of the growth rate in national income or of the rate of capital formation. The new emphasis is on wider, qualitative aspects of development, such as an improvement in income distribution, employment, health, housing, education and so forth.

One fact to emerge clearly is that an increase in economic growth does not necessarily benefit all sections of society. The conventional wisdom was that an increase in national income would somehow filter or 'trickle down' to the underprivileged. For the most part this has not happened. This is partly because the growth process itself does not necessarily bring about changes in the structure of society. By abandoning the narrow 'sectoral' approaches to development and promoting instead structural changes, nations will be able to bring about improvements in income disparities, unemployment, housing, nutrition and health standards.

Third is the relationship between our understanding of the interaction between man and nature and the design of development policies and objectives. Interaction between forces of change, namely resources, environment, people and development, make it necessary for governments to think in terms of trade-offs between different courses of action. Development objectives must be carefully evaluated in order to identify the environmental consequences of each action.

It is imperative to avoid the actions of one sector having harmful consequences on other sectors. For example, say an energy department builds a hydroelectric dam, but the forestry department does not follow a sound watershed conservation policy. The result will be a silting up of the dam and a failure to meet the projected energy production targets. A 1979 report by the United States Agency for International Development (USAID) observes that deforestation in northern Luzon in the Philippines has silted up the reservoir of the Ambuklao Dam so fast that its useful life has been reduced from 60 to 32 years. This is only one example, there are many more.

We need to devise policy frameworks to avoid these kinds of problems. This requires both an assessment and an evaluation of environmental impacts. Both assessment and evaluation present different problems. For sound environmental management, it is necessary to develop tools of analysis and methodologies that will assist in decision making on the basis of more complete information on the full economic, social and environmental benefits of development policies. We need to promote systems analyses which can show that actions considered of little importance in terms of one sector, such as watershed management or pollution control, often have substantial contributions to make to the attainment of environmental goals. Although UNEP is involved in programmes related to environmental impact analysis, social cost–benefit analysis and integrated physical planning, much still remains to be done in this field.

There are compelling reasons to proceed with urgency. There is mounting evidence that excessive demands are being made on limited resources, and the carrying capacity of fragile ecosystems and the overexploitation of natural resources cannot be sustained. The industrialized nations rely on the developing countries as the source of their raw materials, but already the renewable resource base in many of those countries has been seriously undermined. Deforestation, desertification and soil erosion are only a sample of the outward signs. The still commonly held belief that there is a conflict between development in its wider, qualitative aspects, or even economic growth in its narrower aspects, and environmental quality is an illusion. A harmonization of interests can be achieved through a rational choice of options.

While previous programmes, research and other activities have helped industrialized nations such as Japan to deal with the environmental problems that emerged over the last decade and to develop useful guidelines for environmental policy, the advent of the global recession today calls for a new, more integrated approach to environmental protection.

There are two aspects of such an approach. A realistic assessment of the costs of environmental protection is needed. To this end, UNEP is developing a programme of environmental cost–benefit analysis, begun two years ago. The first results of that study indicate that an investment in environmental protection pays immediate dividends.

Businessmen tend to overestimate the costs of environmental regulations, and environmentalists to overestimate the benefits. We need to see the situation in its proper perspective. Such a perspective indicates that environmental protection measures yield significant economic and social benefits. This is not only in respect of gains in mortality rates and improvements in working conditions, but also in such specific areas as the productivity and profitability of investment, technical innovations, energy use and increased amenities.

In the majority of OECD nations – most notably in Japan, Norway and FR Germany – the level of gross national product is found to be higher and growth marginally faster with existing environmental programmes than without them. What emerges clearly is that capital expenditures in the environment sector could make a useful contribution to GNP growth rates.

Savings and benefits can be made across the board. For example, the results of a French study on 24 pollutants revealed that the cost of pollution in 1978 came to approximately 4% of the GNP. Cost–benefit analysis indicated that an investment in environmental protection of no more than 2% of the GNP could have saved those additional costs. Similar studies have produced parallel results in Italy and the UK.

In 1978, total pollution costs in the USA were estimated to amount to $47.6 million. Studies carried out for the US Government's Environmental Protection Agency showed that health benefits alone from control of pesticides have amounted so far to an incredible $8 billion per year. Potential benefits are estimated to be as high as $41 billion annually if a 60% reduction in air pollution particles is achieved.

When we look at inflationary pressures and the effect on employment levels, we are confronted with an interesting situation. While it is true that the application of the 'polluters pay principle' – PPP for short – has led to a small increase in the rate of price rises, this is offset, firstly, by the fact that PPP is designed to bring about appropriate changes in production and consumption and, secondly, by the fact that the final price increase is always far below the rate of inflation. Over a period of five years the largest rise on this account was in Japan which, even so, amounted to a mere 3.8% for the entire period.

Expenditure on environmental protection also creates new jobs. A study carried out by the US National Academy of Sciences estimated that there were nearly 7 million people employed directly in pollution abatement by the mid-1970s. In France another study indicated that, as a direct result of environmental policies, 370 000 new jobs were created in 1979 alone. Today, in the USA, France and FR Germany, 1% of the total workforce is employed in environmental jobs.

It would clearly pay governments to coordinate energy and raw materials policies with environmental policy in order to secure overall savings through the recovery of wastes. These are net gains in addition to the gain in environmental quality as such.

Pollution is largely a function of waste, and private companies are finding that they can make large profits by retrieving pollutants. UNEP has discovered that in Switzerland, the USA, the UK and FR Germany a number of industries have increased their profits, in some instances by as much as 40%, by recycling wastes which would otherwise have been released into the environment.

Japanese industry is a world leader in this field. The Japanese cement industry is currently using 35 million used car tyres per year in the cement making process. By burning these tyres 2 million kilolitres of imported crude oil are being saved every year by the cement industry.

In the field of human settlements, environmental planning has produced a reduction in the cost of urban design and control. It includes the preservation of old buildings and community centres instead of demolishing and building anew. Renovation is not necessarily more costly and its resource content is different; it uses more labour, less energy, fewer raw materials and less capital. Returning to the issue of employment, a recent US Federal Government paper estimated that the installation of ceiling insulation, automatic thermostats and furnaces in 34 000 homes would produce 480 000 new jobs over a two-year period. This would be in addition to the savings on the cost of importing oil.

There is no evidence to suggest that the foreign trade of any country or its balance of payments have been significantly affected as a result of the introduction of environmental protection measures. Environmental expenditures have generally been absorbed at the plant and industry level, as in the case of Japan, and not passed on in export prices.

Similarly, specific studies indicate that the costs of environmental protection measures are susceptible to proper direction and development. The control costs, even in high polluting industries such as iron and steel, pulp and paper, aluminium and so on, do not exceed a few percentage points of the average product price. Control costs vary from country to country and between different regions but, on the whole, have not significantly distorted trade.

One positive element to which not enough attention is paid is the impact of environmental measures, and of problems that such measures create for industry, on rapid technological innovations and increases in productivity. In Japan's case, while it is true that the environmental protection measures have been introduced at a cost higher than in most other industrialized nations, they have not materially affected the competitiveness of its industry. In fact in many cases it has made it more competitive.

The outstanding example is the Japanese car industry. The strict emission control measures introduced in the early 1970s anticipated the trend in other countries, with the result that Japanese low fuel consumption and low-polluting cars are now in demand throughout the world.

The preliminary findings of a study carried out by the ILO have revealed that the introduction of equally stringent pollution controls in other sectors have encouraged Japanese industry to introduce cleaner technology and to renovate the older polluting and wasteful plants. As a result of this, the ILO study points out, Japanese industry is now cleaner and more efficient than that of many of its foreign competitors.

But even supposing that environmental protection measures did make industry less profitable and led to a slow-down in the rate of growth of the GNP they would still be enhancing social welfare. This realization prompted the OECD environment ministers to conclude collectively in 1980 that 'In the long run environmental protection and economic development are not only compatible but interdependent and mutually reinforcing'.

The second aspect of the new approach relates to policy planning. We are living in a period of rapid change. We are facing problems posed by an unstable economic situation and an increasingly complex society. Scientific and technological developments have opened up unprecedented opportunities, as well as new dangers and problems. A wide range of new social objectives need to be identified and defined. Failure to take account of popular will in areas which affect public welfare might weaken political systems. On the other hand, greater demands and broader participation must necessarily make the planning process more complex and less amenable to simplified approaches.

In this situation we require, as never before, a clear understanding of the interrelationship between people, resources, environment and development. New ways have to be sought to overcome the complex demands which narrow-focus sectoral planning cannot meet. A more global planning process needs to be employed and it must be both integrated and continuous.

The practical problem arises from the fact that, although methodological problems have not all be solved, it is becoming increasingly urgent to incorporate environmental considerations into a decision-making process for which integrated planning provides the only vehicle. There are, for instance, a number of major economic and investment decisions pending before Japan. I refer to decisions which have both short- and long-term effects in areas such as the future of urban settlements, food production, industrial growth, transport and the role of heavy industries. Difficult and wide-ranging decisions will also have to be made in the energy field, including decisions about exploiting appropriate sources of renewable energy, new sites for power stations and energy conservation.

Japan is faced right now with the challenges of integrated planning, which involve effecting a series of evaluations of data and information to secure set objectives. In order to succeed with this daunting task, it will be necessary for Japan, firstly, to adopt anticipatory policies for integration of environmental considerations in economic and social decision making, secondly, to accept the improvement in the environmental quality of life as a basic element of its social policy and, thirdly, to encourage popular participation in the resolution of conflicts and trade-offs between different options which are available.

It is evident that these tasks cannot be accomplished alone, either by UNEP at the international level or by Japan's environmental machinery at the national level. The people in government, in industry and in the academic field must all come together to identify a new way of thinking about systemic relationships and determine means to manage effectively the processes of social growth and change.

It is also evident that the tasks are urgent. UNEP fully endorses Japan's summing up of the problem, in a report to the Prime Minister, when it is said that, despite uncertainties and differences as to details, evidence shows clearly the urgency of these problems taken as a whole. As the report went on to say: 'This group believes that there is a substantial danger of being too late if steps are not taken quickly'. Let us move now.

3

Safeguarding the environment

Statement to the Session of a Special Character
of the Governing Council
of the United Nations Environment Programme

Nairobi, Kenya, May 1982

This Session of a Special Character, to which all member states of the United Nations have been invited to commemorate the UN Conference on the Human Environment, is the outcome of a General Assembly resolution. As stated by the Secretary General, the Assembly endorsed the concern of UNEP's Governing Council that the commitment to the environment made by nations attending the Stockholm Conference should not be allowed to fade. The task facing us is to give new impetus and a new direction to the environment movement for the next 10 years. The developments of the last 10 years will be reviewed and a new course will be charted that takes account of the successes and failures.

The option facing governments at this time is stark: take action or face certain disaster. Ten years ago the options facing the 113 nations attending the Stockholm Conference were not so obvious. Nevertheless they agreed a sweeping action plan 'to safeguard and enhance the environment for present and future generations of man'. The Stockholm conference also recognized that action would need to be accompanied by more knowledge of our environment.

The record of achievements in implementing the Stockholm Action Plan has been documented. A summary of the main environmental developments of the past 10 years and a statement of the actions required is laid out in the 'Retrospect and Prospect' document. This single document is the product of an exhaustive consultation process involving governments, UN agencies, the scientific community and environmental non-governmental organizations. It reflects the broad agreement that now exists among nations on how we can solve our common environmental problems.

Developing environmental knowledge

There has been progress since Stockholm, mostly in increasing our knowledge of the environment. Ecology and environmental sciences have matured in the last 10 years. In some cases, increasing knowledge has enabled us to turn the theories of a decade ago into fact, in others to dismiss them entirely or to uncover new areas of concern. For example, desertification was scarcely mentioned at Stockholm; today, with arable and grazing land being turned into desert at the rate of 6 million hectares per year, it is seen as one of the most devastating environmental problems.

A subtle change in emphasis took place during the decade, from worrying about the changes in the state of the physical environment to concern over the causes and impacts of such changes. Throughout the decade our perceptions and our understanding have continuously evolved. Ten years ago preserving wild plants and animals was seen as a worthwhile activity in itself. Now there is widespread recognition that the future of agriculture, and of pharmaceutical and other industries, hinges on the conservation of wild species.

This new view of wild species as a resource was paralleled in other areas. Increasingly, as the decade proceeded, we came to regard forests, soil, fish, clean air and fresh water as resources to be conserved. By the end of the decade this is no longer an activity that is vaguely desirable, but one crucial to our future survival.

During the decade we also came to realize that in the environment everything is related to everything else. Atmospheric ozone depletion, possible climatic warming due to the build up of carbon dioxide, acid rains and nitrogen shortages in the soil were once seen as separate problems, soluble on their own. We now know they are closely linked through the global cycles of carbon, oxygen, nitrogen and sulphur.

Harsh experience has shown that environmental neglect in one quarter can have harmful consequences elsewhere. The useful life of some dams, for example, has been halved by siltation caused by unchecked deforestation in watersheds.

The consequences of environmental neglect can have repercussions far beyond national borders. Oil spillages are a problem for all nations sharing a common sea; acid rain has become a serious problem, not only in the polluting countries but even more so in those countries with the misfortune to be downwind.

During the decade we also noted that some ecosystems displayed remarkable resilience to our transgressions. Yet they also displayed their limits. Loss of grasslands and forests, unwanted changes in freshwater systems and declines in productivity of coastal waters were among the penalties we paid for failing to manage these ecosystems properly.

Poverty is the worst form of pollution. This phrase was coined for the first time at Stockholm. As the decade wore on, we saw how poverty forced villagers and slum dwellers to destroy renewable resources, like fuelwood and soil, on which their future survival depended. In 1982 it is underdevelopment which is still the principal cause of the environmental problems we have to solve.

Stockholm accepted the idea that the solution lay in environment-based development, which enhances rather than damages the environment. Then, it was a revolutionary concept; today, it is common currency among decision makers. Strategies, action plans and programmes have been written with guidelines on how nations can institute a development process which will meet the material needs of their people, while at the same time protecting their environment.

A further step came with the publication of the World Conservation Strategy; it provided decision makers with an overall blueprint for exploiting their interconnected environment in a sustainable way. A remark made by a leading minister on the day the strategy was launched was significant: he said, 'in any individual decision the starting point will be to conserve what matters – those who have a contrary objective must bear the onus of proof'.

Consequences of poor environmental management

Unhappily, governments have not matched this developing environmental knowledge with deeds. The concepts for ecologically sound management have been imperfectly or too slowly applied. In some cases they have been ignored entirely. The inevitable consequence is that the fundamental objective of Stockholm, to protect and enhance our environment for future generations, has not been fulfilled. On virtually every front there has been a marked deterioration in the quality of our shared environment. The result is that now, when we need *more* of everything – more housing, more shelter, more food, more jobs, more fresh water – the planet's capacity to meet those needs is being undermined.

This means that our room for manoeuvre has narrowed considerably since 1972. Food production, despite advances in agricultural methods, has barely kept pace with the increase in population. In Africa it has actually fallen behind. Circumstances will become bleaker still over the next decade if we continue to permit farm and grazing land to be reduced to zero productivity at the current rate of 20 million hectares per year.

Today, 450 million people are chronically underfed. Millions will be added each year unless we stop the haemorrhage of soil loss. And we *can* stop it. We have an action plan to stop desertification, we have soil conservation strategies and tree planting programmes: widely implemented, these strategies will preserve the one third of the world's arable land now at risk.

Tropical forests are being depleted at a rate close to 12 million hectares per year, and disappearing with them are their precious mines of irreplaceable genetic resources. Like forests and soil, our freshwater resources need to be managed properly. If energy was the source issue that attracted most attention in the 1970s, as the 1980s unfold water is likely to take its place alongside. Demand for already overstretched freshwater supplies will increase; even now some West Asian cities are being forced to withdraw fossil reserves to meet expanding needs. Pollution and conflicting demands on surface and ground sources could lead to friction between nations.

The need is for water not only in more plentiful supply, but also for water that is clean. A large proportion of diseases affecting people in the developing world are waterborne. Today, one in four people living in the cities of the developing countries has no access to clean water. In their rural areas, where the majority still live, the situation is even worse – more than 70% must drink and wash in dirty water.

People's health is also being put at serious risk by the increasing volume and numbers of potentially dangerous chemicals released every day into their environment. A similar threat is posed by the transport and disposal of hazardous wastes, especially long-lived radioactive waste.

Increased environmental awareness among decision makers

These are only samples of the environmental problems we are carrying over from the previous decade. Others will emerge as we approach the turn of the century. The magnitude of the problems we are facing can not allow lack-lustre performance on the scale we have seen over the past 10 years.

I believe that governments – singly and collectively – will respond during the next decade in a more serious way. This belief is based on a number of very positive developments. Most notable is a remarkable expansion of environmental awareness in decision-making circles.

Government environmental machineries have expanded in numbers from 10 to 106. These machineries have also expanded in influence and power. This growth has been accompanied by an increase of more than 5000 in the number of NGOs concerned with resource issues.

We have also seen the start of a new era of cooperation between nations to safeguard shared resources. Regional seas treaties have been approved, conservation treaties ratified, antidumping conventions observed and river basin commissions established. These developments are ample reason for confidence.

A major landmark was the commitment of the world's leading international development assistance institutions – which, between them disburse more than 14 billion dollars each year – to funding only sustainable projects. 'Sustainable development, and wise conservation are, in the end, mutually reinforcing – and absolutely inseparable goals' concluded World Bank President Clausen in a recent major public statement. Several bilateral aid agencies, including those of the USA, Sweden and FR Germany have made similar commitments. Others are starting to follow their lead.

Benefits of environmental protection

There would be little cause for confidence in the circumstances of the current recession if there were no facts and figures to show that environmental protection pays, and in hard cash terms. Several government surveys have shown, conclusively, that an investment in environmental protection can result in savings of up to 2% of the GNP.

In both developed and developing countries, environment policies have created new jobs and industries. In France 370 000 new jobs have been created by environmental protection. According to the Organisation for Economic Cooperation and Development (OECD), strict controls introduced by the Japanese Government have been a major stimulus to growth. There is no evidence to show that environmental protection has been causing any significant inflation. The highest price rise induced by environmental regulation was recorded in Japan, and there, during the five-year period after 1974, it was just 0.7% per year.

Some sectors of industry in countries as diverse as PR China, Belize, Brazil and the USA have been discovering on their own initiative that productivity can be boosted by retrieving pollutants and using them as resources. UNEP has found in some cases that profits of up to 40% have been made through recycling.

One important element seems to have attracted no attention during the decade. Nations have made little effort to introduce an environmental accounting system. If oil, industrial plant and so on are counted among a country's assets, why not surface soils, clean air and water, and gene pools?

It can take nature a millennium to accumulate soil to a spade's depth, yet it can be removed in a matter of months. That reserve of soil is more precious than any gold reserve, but we do not count it so.

Were an environmental audit system to be applied, I am certain greater care would be taken to correct what has been described recently as 'biological deficit financing'.

The case for environmental protection does not rest here. Though the health, spiritual and aesthetic benefits of environmental protection can never be subjected to accounting in strict monetary terms, they are invaluable and lasting. The case for environmental protection rests ultimately with global peace. Despite overwhelming evidence, there is little recognition in government circles of the extent to which resource exhaustion contributes to economic recession and threatens security. We need to widen the definition of national security to take account of global environmental risks that are unknown to traditional diplomacy and indifferent to military force.

Gone are the days when a nation, no matter how powerful, can consider itself immune to disturbances or disruptions of an environmental nature occurring elsewhere. The Brandt report has already said: 'Few threats to global peace and survival of the human community are greater than those posed by the prospects of a cumulative and irreversible degradation of the biosphere on which all human life depends'.

As the environment is interdependent, so are nations, whatever their political complexion or stage of economic development. It therefore serves the self-interest of the industrialized and other rich nations to invest in the environmental security of the poor, developing countries. A similar conclusion was to the fore in a recent report issued by OECD which said that 'The destabilisation of the world's ecosystem and the degradation of environmental systems in particular regions are among the fundamental problems which create actual or potential risks of concern to both developed and developing nations'.

Implementation of environmental action

In the Retrospect and Prospect document we have distilled the recommendations for action required in order to head off the gathering environmental crisis. These suggestions are not UNEP's alone, nor are they to be implemented by UNEP alone; they are to be undertaken by the whole UN system and above all, by governments themselves.

Implementation of those actions to secure all our futures will require a major transfer of resources. The water and sanitation decade needs 300 billion dollars to be committed between now and 1990 to meet its target of clean water for all. The action plan to combat spreading deserts requires 1.8 billion dollars to be committed yearly until the end of the century. A great deal of money must be spent to make up for the years of inaction. However, it represents only about 5% of the sum nations now spend on arms.

Conclusion

To recap, it can be said with confidence that there is a widespread recognition in governments, international bodies, and now increasingly in industrial circles, that it pays in both the long and the short term to invest in environmental protection as the key to sustainable development.

We also have the tools to translate concepts into actions. But will these changes in our understanding of the importance of the environment be accompanied by political and resource commitments? And how can we ensure that our expanded, and expanding, knowledge of the environment is converted into better decision making? These are the twin challenges facing this Session of a Special Character.

In addressing these challenges it must be kept in mind that, within the limits of the brief handed to us by the Stockholm Conference, UNEP's role is that of a coordinator and catalyst. Our job is to encourage and persuade others to take action. In fulfilling the task of coordinating government action, UNEP depends upon the policy guidance emanating from governments themselves during the regular sessions of its Governing Council. Within the UN system we have various effective mechanisms and procedures for coordination, the latest being thematic joint programming and the development of the System Wide Medium Term Environment Programme. With non-governmental organizations and the scientific community we have several channels of communication; we are always trying to improve them.

We attempt to maximize the impact of our catalytic function by using our limited financial resources to instigate necessary actions and mobilize more resources. For example, with our Global Environmental Monitoring System we coordinate a number of monitoring networks organized by an array of specialized agencies to which they and governments contribute much larger funding than we do ourselves. Another example is regional seas where we provide seed funding to get governments together to work out action plans. After these plans are agreed, governments gradually take over the responsibility for implementation.

There are a number of ways in which the catalytic role can be perfected and we are working on them. But the crucial factor in improving UNEP's difficult function is to increase the resources available to the Environment Fund. Uncertainties over the precise amount and phasing of voluntary contributions to the Environment Fund have adversely affected the performance of this organization.

In recent years, while demands have been increasing dramatically, contributions have been decreasing in real terms. The 31 million dollars we received last year are equivalent to 16 million dollars in 1973 – in constant monetary values. This is over 10% less than we received in 1974. At the end of last year we had to cut back almost half of the Programme activities across the board. In some cases we found we simply could not honour our commitments to governments and supporting organizations.

No organization can be expected to plan effectively when it does not know when or how much funding will be provided. This is a crippling factor. I appeal to governments to consider its seriousness.

We face today a problem which no previous generation has had to face. In 1982, nations have two choices: to carry on as they are and face, by the turn of the century, an environmental catastrophe which will witness devastation as complete and as irreversible as any nuclear holocaust, or to begin now in earnest a cooperative effort to use the world's resources rationally and fairly.

We hold more firmly than ever to the Stockholm principle that of all things on earth people are the most precious. There is still great potential to meet the needs of the people alive today and the nearly two billion additional passengers we will see come aboard spaceship earth between now and the end of the century. We can solve our environmental crisis, but first governments and their people must show they have the will to do so.

4

Disarmament and environment

Statement to the United Nations General Assembly
Special Session Devoted to Disarmament

New York, USA, June 1982

Many eloquent arguments have been made in favour of disarmament in the past, but the fact remains that in 1982 we are still where we were in 1978, facing the same stark choice: survival or annihilation.

Modern weapons have the deadly capacity not merely to wipe out cities, industries, even entire populations, but to destroy the life-giving systems on which we all depend. The need to preserve our shared environment provides the most persuasive argument for nations to stop the dangerously escalating spiral of the arms race.

The peoples of the world do not want war; they want the arms race to be reversed and they want the danger of nuclear war to be eliminated. They are puzzled and bemused. They see obvious contradictions in the attitude of the world community to the whole question of military activity. On the one hand, the numerous conventions, treaties and agreements provide clear evidence of a widespread desire to prevent the more devastating forms of warfare. On the other hand, the evidence of mounting military expenditure around the world implies a lack of conviction in the practicability of disarmament, or even of holding forces and arsenals at constant size. The peoples of the world look to the United Nations to show how we can achieve the security which the arms build-up has not provided. The message, though expressed in a multitude of ways and in many languages, is to act now.

During the Session of a Special Character at Nairobi to mark the tenth anniversary of the Stockholm Conference on the Human Environment, 105 governments were at one in declaring that 'the human environment would greatly benefit from an international atmosphere of peace and security, free from the threats of any wars, especially nuclear war, and the waste of intellectual and natural resources on armaments'. By a special resolution, the nations assembled in Nairobi appealed to governments and the world community as a whole to do their utmost to halt the arms race and thereby prevent a major threat to the environment.

Legacies of wars

Recent UNEP studies have focused on the past, present and future threats posed by weapons and warfare to the complex systems of the biosphere. Wars of the past have had both direct and indirect effects on the environment. They changed agriculture, shifted the margins of deserts and grossly disturbed the balance of ecosystems. The second world war caused a reduction in agricultural productivity of 38% in 10 nations. A modern day conflagaration would have immeasurably more effect.

A legacy of past wars remains among us in the form of unexploded mines, bombs and shells which endanger those who disturb them, make land unsafe to farm or develop for other purposes and hinder mineral exploitation. One government has reported that it has cleared more than 14 million land mines left behind after the second world war, and that clearance was continuing at the rate of 300 000–400 000 mines per year. These remnants of war had killed in the same country close to 4000 people, most of them children, and had injured more than 8000, of whom about 7000 were children.

Chemical and biological weapons involve deliberate pollution by the release of toxic chemicals or harmful microorganisms. Chemical deforestation in fragile tropical or semiarid areas could create rapid erosion and irreversible desertification. Chemical, bacteriological or bilogical weapons disturb agriculture and ecological balance for a long period of time. In Vietnam, chemical herbicides completely destroyed $1500\,km^2$ of mangrove forest and caused some damage to a further 15000 km^2; natural recovery is proceeding at a disturbingly slow rate.

Tragic though such cases are, it was the widespread social disruption and uprooting of large numbers of people that created the most persistent human effects of past wars. About 17 million were displaced in the Second Indo-China War; such refugees not only experience personal suffering and economic loss, but exert severe pressures on the environment in the areas to which they migrate.

The disposal of obsolete weapon stocks poses threats to the environment, and the growing volume and increasingly destructive capacity of the world's weapons constitute an obvious risk to be biosphere and the people living within it. Environmental manipulation has added a new dimension to our self-destructive capacity. Today, the threat of a nuclear holocaust forces us to take a long, hard look at the human and environmental consequences of an event that must never be allowed to occur.

Environmental effects of nuclear war

There is talk in some military circles that a nation can 'win' a nuclear war. This is an unrealistic and dangerous notion. A full-scale nuclear war would, among other things, destroy all major cities in the northern hemisphere, killing the bulk of the urban population there by blast and fire, and the bulk of rural population by radiation. Many millions in the faraway southern hemisphere would also be killed by radiation from fallout. Though relatively unpredictable, the long-term consequences could affect the global climate, reduce the ozone layer and induce serious genetic effects. Use of nuclear weapons in a full-scale war would destroy vegetation and lead to soil erosion over vast areas. Ecological recovery in such eroded areas would certainly be extremely slow.

A recent study has indicated that, if a nuclear war took place at the beginning of the growing season, food production would be almost totally eliminated in the northern hemisphere. If sunlight was blocked out by light-absorbing particles lofted up into and spread around the atmosphere, a large amount of plankton – which provides the basis for most marine life – might die in about half of the northern hemisphere. In these grim circumstances, it would be difficult to conceive of continued survival for those who remained alive in the years following such a war.

Environmental damage resulting from the arms race

Even if not one single bullet was to be fired in anger, the arms race would still be causing damage to the environment. The Stockholm Conference recognized, and the Nairobi conference endorsed the view that the gravest threat to our environment is posed by poverty resulting from underdevelopment. The way to alleviate poverty is through sustainable development. But this objective will stay elusive while military activities continue to jeopardize the process of development by diverting substantial amounts of human, natural and financial resources to arms production. Endless examples can be given but let me describe just one. An estimate has suggested that military research and development absorb scientific and technological capabilities 10 times as great as are available in all the developing countries.

There may be room for argument over the details behind the figures of military expenditure, and over the general relationship between such expenditure and other national investment. What is not challengeable, however, is the fact that the increase in military expenditure takes place at a time when more than 70% of the peasants in developing countries lack drinking water supply services and more than 85% of them have no effective sanitation; 450 million are chronically undernourished, 1500 million people have no medical services and 250 million under the age of 14 do not attend school.

These and a host of other appalling facts can be enumerated. On the basis of UNEP's work in this area, some concrete recommendations have been formulated as a first step to safeguard humanity and to protect the human environment against arms production, stockpiling and use. Eight of them stand out. These are:

1. continuing assessment, monitoring and evaluation of the impacts of military activity on the environment;
2. encouraging studies on relationship between security and stability of ecosystems at local, national and wider levels;
3. demilitarizing ecologically important regions;
4. ensuring that outer space is not used for hostile purposes;
5. a call upon all powers that have not yet ratified the treaty banning nuclear weapons tests in the atmosphere, in outer space and under water, to do so;
6. development of treaty by which all nations would pledge themselves not to be the first to use nuclear weapons in warfare;
7. inviting the Conference convened to review the Convention on the prohibition of military or any other hostile use of environmental modification techniques envisaged for 1983 to consider the possibility of banning antiplant chemical warfare and of strengthening the Convention itself to prohibit all hostile uses of environmental manipulation techniques;
8. establishing a ban on any weapon or technique, existing or potential, which would devastate a wide area and threaten the regional or local ecological balance.

We should not underestimate the forces which militate against disarmament: the insecurity nations feel, the influence of the military/industrial complexes, the ambitions of individuals and so on. But, nor should we underestimate the powerful thrust for peace which, like the environment, knows no political boundaries and which looks to the United Nations to give it form and substance.

It is also true that the peoples of our global family of nations have seen that expenditure on arms does not bring security. It is also a fact that, despite overwhelming evidence, there is little recognition in government circles of the extent to which resource exhaustion contributes to economic recession and threatens security.

The way ahead

The question must thus be asked – are we nearing the end-point of our evolution? Have the means become all-powerful, and the ends obscure? Or is there now developing an affirmation of love of life and the earth whose fate we share? Never before in our history has there been a decision of this magnitude to be made.

We have arrived at that point in time where the hand of mankind – which all too often has been the wielder of death and destruction – might yet be transformed into the healing hand, with an ethical responsibility to prevent further damage from occurring, and to heal an already wounded planet. With a reversal of priorities, it would become possible to deal with the appalling poverty, scarcities and environment degradation so prevalent in the world today. Real security can only come when nations cooperate to exploit equitably, wisely and sustainably this planet's dwindling treasure of natural resources.

5

Environmental information for engineers

Closing statement to the Seminar on Environmental Information for Engineers

Paris, France, November 1982

This Seminar on Environmental Information for Engineers has provided a forum for the exchange of experience in the handling of information systems in a large number of countries, as well as the opportunity for greater cooperation at the national, regional and international levels. It has also identified the needs of a specific user group, the engineers, with all the ramifications and impacts of their profession on the whole process of development in every corner of the world and the major role they have in influencing the quality of life. The recommendations of the seminar and the round-table discussions will have an important impact on UNEP's plans and activities.

Information dissemination

The topic of environmental information and its dissemination is, as you know, a vital component of UNEP's work. If environment and development are to be compatible, then information dissemination is the channel through which knowledge and experience are transferred to link the two.

Ignorance is one of the most important causes of environmental impoverishment. With an adequate system of information we can learn not only not to repeat the mistakes of others, but also to benefit from their achievements. In this area of environmental information the only acceptable operating philosophy is 'share and share alike'.

Information is a unique resource. While on the one hand it can be expensive to acquire, compile and collate information, it is not like other resources which, once dispersed, are used up and lost to their source. The acquirer of information becomes its new possessor who, in turn, can share it with others without losing it. Such multiplying and catalyzing effects are therefore significant. There is, of course, the recurring issue of confidentiality – I will return to this aspect later on.

Information and data

As has been pointed out during this round-table discussion, information and data are two terms often used synonymously. They are, however, disparate, with different implications and I would like to share with you my perception of each. Information is qualitative and can be subjective. It is a product of observation, perception,

25

intuition, circumstantial evidence and value judgements. It is nevertheless an indispensable part of the decision-making process.

Data, on the other hand, are measurable, quantitative, based on numerical values and present, ideally, an objective account of the facts. The element of subjectivity enters when data are interpreted to relay their information content.

Both environmental information and data have important roles in promoting public awareness and decision making. They support and augment each other. However, their distinct roles and implications need to be recognized and understood clearly.

The information society

The sometimes bewildering advances and developments in computerized information systems have brought the industrialized nations to the brink of what some have begun to describe as the 'post-industrial era of the information society'.

An OECD study has reported that profound structural change is occurring in the economies of its member countries relating to the growth of information activities, and to the increasing share of information goods and services in gross domestic product and international trade, which conventional statistics do not reveal. In the USA, the information industry is growing at twice the rate of industry as a whole and accounts for a greater percentage of the US gross national product than the manufacture of tangible goods.

The OECD study also reported that the personnel employed in all aspects of the burgeoning information industry have been growing at the rate of nearly 3% over five-year periods in the OECD countries. Information-related occupations now account for more than one-third of the labour force. The report found that 75% of the growth in information occupation is in the service sector, the remainder in the manufacturing sector.

Information systems

The recent growth in environment information and information dissemination systems has been staggering. For example, 10 years ago 180 scientific and technical journals relating to environment and conservation were published, now it is over 1100. A huge number of information bases are also now in operation, serving national, regional and international needs. One convenient indicator is UNEP's INFOTERRA network which involves 116 countries and nearly 10 000 sources.

We can expect an additional impetus to growth to come from the information needed on environmental accounting. Environment cost–benefit analysis is maturing rapidly and there is a clear need to construct systems to meet the demand for information on what is perhaps the most important aspect of environment protection.

The growth in the quantity of environmental data has kept pace with the developments in information systems. Many countries are now systematically collecting and compiling environmental statistics. In Poland, for example, the monitoring system envisaged for air quality alone will utilize 72 automatic stations and approximately 800 observing sites. Data gathering by satellites is providing us with a better picture of the changing state of our environment. Satellite images are also providing us with an update on the status of natural resources, such as forest cover and groundwater supply.

Within UNEP, the Global Environmental Monitoring System (GEMS) and the International Register of Potentially Toxic Chemicals are two activities that are systematically collecting and evaluating data. There has been a joint effort by various members of the UN system to produce the required environmental information. UNEP is the beneficiary of the mines of information available to each member of the UN system, particularly the specialized agencies.

The practical applications of satellite data gathering are legion. For example, the data collected by the WMO-coordinated World Weather Watch programme is proving to be of great value as a navigational aid. The collected data can also be used to minimize the effects of hurricanes and floods and other natural disasters. Yet another example is the present network of 70 monitoring stations in 20 countries collaborating in a UN-sponsored exercise in Europe to measure the long-range transport of SO_2 and the impacts of acid rain.

With this surge in environmental information and data showing no signs of abating, a number of policy issues are emerging. Many of these were again identified during the round-table discussion:

1. The need for quality to accompany increasing quantity. Systematic assessment, evaluation and updating are the requirements for quality. Related to this issue is relevance. There is a need for systems which will provide information and data relevant to a particular country and/or region.
2. The need for compatibility in methods for analysing and reporting of data, so that the results will improve the chances of acceptance and be amenable to comparison. A good example is testing for chemical safety which can cost more than half a million dollars. If the test results obtained in one country are not accepted in another and have to be repeated, then this will become a prohibitively expensive procedure. The International Programme for Chemical Safety, spearheaded by the WHO and carried out in cooperation with ILO and UNEP, is promoting harmonized laboratory practices and analytical methods to ensure that data will be internationally comparable and acceptable. The work of the International Standards Organization and the OECD, with its testing protocols, has the same objectives.
3. A requirement for information and data to be in a form manageable and appropriate for the user.
4. A need for data to be presented in a manner which is jargon-free and readily comprehensible.
5. A requirement for environmental statistics to be representative. The many-sided aspects of environmental change, together with the sheer quantity of data, conspire to make complete coverage neither practical nor feasible. To aid decision making, data collection techniques need to be designed to permit the compilation of representative statistics.
6. The need for accessibility. The high cost of equipment and services, and the sophistication necessary to use the system successfully, could impede access by some countries, leading to a division between the 'information rich' and 'information poor'. An appropriate balance has to be struck between the need for equal opportunity of access and the costs of access.
7. The need for user protection. Appropriate controls are required to ensure that, first, what has been purchased matches with what was advertised and, second, that an unsatisfactory information product can be returned.

8. The need for increased education and training, particularly in developing countries, in the application and use of computerized information/data systems.
9. Another policy issue concerns the matter of confidentiality referred to earlier. In meeting the legitimate need to protect the developer's investment in his own information and data base for manufacturing a product, the welfare of the general public, as well as of the global commons, must be taken into account. The issue centres around who should make the decision on whether a certain type of information should or should not be divulged, as well as what should be the procedures to be used in this decision-making process. Public concern must be taken into account, and the public has an absolute right to be kept informed of any environmental and health risks.
10. The issue of sovereignty is coming to the fore with the use of satellites for remote sensing and gathering of resources data on a particular country, without that country's knowledge or approval.

In the light of these emerging issues, it would seem that international guidelines or rules for information and data networks may soon have to be drawn up in order to bring about harmonious development. This, as well as the promotion of compatible systems, is a promising area for international collaboration.

Engineers and the environment

There is a need to identify the user and the objective carefully. In this connection, I am particularly pleased to note that this seminar was specifically targeted at engineers. It may be a platitude to state that without engineers development cannot take place, but it is one that cannot be repeated too often. The contribution which engineers can make towards enhancing, conserving, protecting and managing the environment and its resources, though crucial, has not received nearly enough attention. Engineers need environmental data in their calculations of waste treatment systems and objective assessments of the impact of pollutants on the environment. Such information and data will assist decisions in choosing cost-effective techniques, and also in ensuring the optimal design and operation of equipment.

In short, provision of relevant and readily comprehensible environmental information and data will help meet our shared objective of sustainable development, which brings in train long-lived profits. To conclude, I would like to wish all engineers success in promoting the use and application of environmental information and data in the engineering profession, so that development can take place in harmony with the environment.

6

Highways to nowhere

Statement to the Berlin Environment Forum

West Berlin, FR Germany, November 1982

This is the 10th Meeting of the Environment Forum. These occasions have provided a unique opportunity both to broaden and deepen the dialogue between environment NGOs and parliamentarians, local government authorities, industry and scientists. These fora have been a testament to the fact that managing our environment is an activity of crucial relevance to all sectors of society.

We are 10 years on from the 1972 Stockholm Conference on the Human Environment – a conference in which the NGO movement played an essential role. A landmark such as a 10-year anniversary provides us with an opportunity to take stock of where we are, and where we should be going. However, first let us imagine we have gone ahead 90 years to the year 2072 AD. What, I wonder, in this centennial year of the Stockholm conference, would the person asked to address such a forum be saying?

My fear is that he or she may be saying something along these lines: 'By the year 2000 the worst case predictions had come to pass. The developed countries failed to heed warning after warning that the environmental crisis in the developing world, which grew steadily more severe, steadily more irreversible in the 1980s and 1990s, was their crisis too. Though the Brandt Report, the World Conservation Strategy, the State of the Environment Reports and many other studies produced after the 1972 Conference had provided a watertight case for global economic and ecological interdependence, a combination of political inertia and preoccupation with the socioeconomic problems of the moment accounted for the continuation of the 20th century global pattern of inequitable and destructive development. Too late came the realization that global economic and political security lay in restoring the delicate, but eminently attainable, balance between people, resources, environment and development. The environmentally induced economic collapse in the under-privileged South resulted in the anarchy and civil strife which eventually caused the global conflagration ...'.

I am hopeful that the year 2000 will see such gloomy predictions confounded – indeed there are a host of recent developments which provide scope for optimism, and I will come to these a little later on.

But we must not be misled by the viewpoint that optimism is the only possible operating philosophy. With a few notable exceptions, the trends are profoundly disturbing. They give an accurate picture of an advancing wasteland which threatens every nation, be it socialist or capitalist or underdeveloped.

● We know, for example, that, mostly through shifting cultivation and firewood collection, the tropical forests are disappearing at the rate of roughly one hectare every two and a half seconds.

29

- We are aware that literally billions of tons of precious, virtually irreplaceable topsoil are being blown or washed away each year.
- We are also aware that more than 20 million hectares of productive land are degraded annually to a state of complete uselessness through desertification.
- We know too that at least 10% of the range of flowering plants, many of which could be of invaluable use to agriculture and industry, are threatened with extinction.
- We know that, through ill planned irrigation projects and a failure to provide a clean water supply and decent sanitation, environmental diseases like malaria and bilharzia are actually becoming more common.
- We are aware that, through siltation, watershed destruction and pollution, the natural drainage systems in the developing countries are being destroyed.

These are only a few of the symptoms of the environmental crisis in the Third World; and let us be clear, this destruction amounts to a process of global asset-stripping.

Environmental mismanagement

As was already becoming known by the time of the Stockholm Conference, the major reason for this asset-stripping was mismanagement which then, as now, is frequently manifested in no management at all. The two fundamental causes are underdevelopment and mishandled wealth.

The poor two-thirds of the world are being forced to destroy the very resources on which their livelihood depends, while the rich minority, in pursuit of wasteful development, is making unsustainable demands on resources, and by so doing is turning its disadvantaged trading partners into proxy victims. And be in no doubt, their economic collapse will drag the developed nations into the abyss with them. It is already happening. While governments of differing political complexions in the Western world try out monetarist prescriptions – the recession obstinately refuses to go away and the queues of the jobless grow longer. It is far from being a matter solely of national concern, for, as authors of the Declaration of Philadelphia over 200 years ago stated, 'poverty anywhere is a threat to prosperity everywhere'.

The markets in the rich world are, in many sectors, saturated; the path to permanent economic recovery lies in the overturn of a global economic system that is not only grossly unfair but which tolerates the undermining of the developing countries' resource base. Throughout the world there is copious evidence of the carrying capacity of systems being overloaded to breaking point. Where life-support systems have already collapsed, the options for the poor are stark: either to flee or stay put and starve.

The refugees from Haiti and from the drought-stricken semi-arid regions of the Sahel are environmental exiles. The well publicized tragic end to the Haitian refugees' exodus has given us a glimpse of a future which does not work; of the consequences of a failure to find the means to manage the soil-and-drainage-protecting forests which just 30 years ago covered fully one half of that country.

The security implications of persevering with the present order have already been identified and cogently expressed in the Brandt Report which stated: 'Our survival depends not only on military balance, but on global co-operation to ensure a prosperity based on equitably shared resources'.

Expenditure on arms

It is the fear of disorder, combined with an unwillingness or inability to think in terms of prevention, that is today leading so many governments to invest in arms, instead of their environment. Though small in global terms, the developing nations' share of total expenditure on arms increased particularly rapidly during the 1970s. This was due in no small measure to the activities of a network of salesmen in the Third World, sometimes in the guise of visiting dignitaries, maintained by the engorged military–industrial complexes of the industrialized nations.

During the 1970s, military resource consumption averaged up to 6% of total output in the rich and poor worlds and up to 28% of central government expenditure. At a time when nations spend one million dollars per minute on arms, we tolerate a world in which development assistance programmes languish in fractions of 1% of GNP; a world in which one person in three has no access to decent medical services, in which 650 million must drink and wash with dirty and contaminated water, and in which nearly half a million children die each month from infectious diseases.

Following the example of the developed world

It is true that a major part of the Third World's economic crisis lies in the uncritical adoption of the North's economic and cultural values. The decision-making elite, for the most part, still receive their higher education in the developed worlds' city-based universities, business institutions and military academies. Often they return to their countries with an unconscious scorn for the traditional, frequently rural-based values; they conceive of economic development mostly in terms of big factories, highways to nowhere, foreign exchange-earning goods and expensively equipped hospitals. These sectors receive the lion's share of scarce investment, while token support only is given to alternative development schemes. In part it is a crisis caused by the belief that the only path to prosperity is to follow the developed world's example.

The idea that prosperity would 'trickle down' to the villager and slum-dweller is now totally discredited. If we are honest, one reason for the present crisis in the Third World is that, through graft and corruption, in several cases manipulated by outsiders, the wealth created has sometimes 'trickled' into numbered bank accounts, and unnecessary shows of wealth.

Solving the environment problem in the Third World

I believe that the current generation of decision makers in both developed and developing countries will be reviled for generations to come, unless, while there is still time, they begin in earnest to break the current cycle of development and resource destruction. And we can break that vicious cycle, for we not only know the problem – we have the solutions to hand.

I am still hopeful that, as the environment component in economic depression becomes more obvious, the prescriptions will be applied. The basis for this hope, for example, lies in the growing support for the international environment and conservation accords; in the undertaking by all the major multilateral development assistance agencies, and now by many bilateral agencies, to support only environmentally sound projects; in the rapidly growing number of Third World governments sponsoring soil and drainage and other conservation strategies; and in the marked willingness of local communities to become involved in resource conservation projects.

I am encouraged too by a growing awareness, stimulated by the NGOs and responsibile media, among the general public in the industrialized countries of the link between the Third World's problems and theirs. The days when a country could consider itself an economic and environmental fortress are gone, and I believe that people – especially the younger generation – are beginning to realize this.

It may be useful to note in this connection that at the end of the 1970s exports from the Western to the developing countries represented one quarter of their total income from trade. Nearly a third of primary products (excluding fuel) came from the developing nations. These transactions are carried out not because of, but rather in spite of the terms of trade which were actually better for the developing nations 20 years ago than they are today.

Governments do not make decisions in a vacuum; demand for action must bubble up from the people, and that demand will not come unless the public are made aware of the environmentally-linked economic dangers which hang like the Sword of Damocles over their future welfare.

The environment movement in the West had its origins in the anger over industrial pollution. Governments are now responding. Just one instance are the EEC's properly rigorous environment directives. Some sectors of industry, too, in the pursuit of profits, are beginning to discover that it makes sound economic sense to recycle pollutants.

What is needed now is a new wave of sustained anger and concern for the many times more grave threats posed by the environment crisis in the Third World. The special job of groups like the Berlin Environment Forum and of the media is to channel that anger to constructive ends. If governments fail to respond, then they have to give their place to those who can. All our futures are at stake and compromises are not acceptable.

Returning to our Stockholm centennial year, I told you what my fear was. My hope is that the speaker at the meeting 90 years hence might say this: 'I believe we can trace the rapid transition from the destructive to the equitable and sustainable development process that took place in the 1980s and 1990s to the 1972 Conference, the centenary of which we commemorate today. For it is in the Stockholm Conference that we see the beginnings of the revolution; it sparked a movement which in the late 1970s and early 1980s provided overwhelming evidence for the need for development which safeguarded the human environment. Responding to a groundswell demand, and motivations of benign self-interest, governments took action in the nick of time to halt the degradation which threatened to destabilize the global community. Thus, by the year 2000 we could see the Stockholm objective of "protecting and enhancing the human environment" being implemented'. Let us hope that, for the generations alive and those yet to be born, this comes to pass.

7

Rolling back the wasteland

Lecture to the India Society
for Promotion of Wastelands Development

New Delhi, India, December 1982

There need be no such thing as a wasteland or waste of any kind. In this new society for Promotion of Wastelands Development, and thousands more environment citizen action groups around the world, I see a developing conviction to tackle the global problem of environmental degradation; a new determination to confront the futile and avoidable wastage of resources.

However, we must recognize that, in a sense, the existence of this Society for Promotion of Wastelands Development and similar groups are a clear indication that we are not yet on the right track. In the 1980s we have the know-how and, more important, the means to prevent the development process creating wastelands and waste products. What we lack is the will to apply that knowledge.

I do not believe there is a single country, developed or developing, which does not need a national programme to revive its degraded lands. With patience, good planning and good management we can begin the vital work of first stopping, then rolling back the advancing wasteland. One of the encouraging features of the previous decade has been the ability of ecosystems to recover from our transgressions. Through proper management, we can nurse our impaired life-support systems back to full productivity. The first priority of all nations is to reverse, now, the process of development which destroys more than it develops. Given their heavy reliance on renewable resources, this message has most relevance for the Third World nations.

A subtle change has taken place in the 10 years since the Stockholm Conference, from worrying about the changes in the state of the physical environment to concern over the causes and impacts of such changes. This planet still has enough intact renewable resources – in effect, our environmental capital – to make prevention rather than cure the top priority. To use a convenient analogy: it is far more cost-effective to tackle the root causes of disease, most notably dirty water, lack of sanitation and poor nourishment, than it is to build expensively equipped hospitals which, at best, can only provide a Pyrrhic cure.

Common to all cultures is the knowledge that 'an ounce of prevention is worth a pound of cure'. With this in mind I have been much impressed by the overwhelming priority The Society for Promotion of Wastelands Development itself attaches to the permanent rehabilitation of wastelands. In my 10 years with the United Nations Environment Programme I have been struck continually by how comparatively easy it is to get new projects started, and how much more difficult it is to keep them going once the novelty wears off and the first burst of enthusiasm fades away. New projects with 'stamina' are needed. In no time at all a rehabilitated wasteland can revert to type.

India's role in tackling environmental problems

The World Conservation Strategy, a plan sponsored by UNEP, the World Wildlife Fund and the International Union for the Conservation of Nature and Natural Resources for the sustainable development of this planet's resources, is but one testament to the unanimity that now exists within the environment movement on how we should go about tackling the processes which create wastelands.

I will endeavour to discuss the causes of wastelands and strategies for their development in their global context. For nearly all aspects of environmental degradation in India are writ large on a world scale. And those problems directly and indirectly affect the environmental and economic welfare of this, the world's largest democracy.

I make an appeal for India, one of the leaders of the Non-Aligned Movement, to do all she can to share her experience in tackling environmental problems with her neighbours in the global South. I am convinced that no action to tackle our common problems will be effective unless nations are prepared to share their experiences. Of course, Environmental TC–DC is a two-way process and India can herself learn much from others. Be assured that UNEP – within the limits of our very narrow resources – stands ready to play the role of 'honest broker' in this process.

Environmentally destructive trends

I intend to concentrate on the four most environmentally destructive trends – deforestation, desertification, soil loss and industrial pollution. Together they have, and still are, turning vast areas of the roughly one tenth of the land surface of this planet that is cultivable into a wasteland. They will continue to do so, at an increased pace, unless we sponsor more intensive action than we have done over the previous decade.

Probably our chief environmental insight of the last 10 years is our dramatically increased knowledge of the interconnectedness of the environment. The causes of destructive trends, and likewise their solutions, cannot be seen in isolation. However, we are still burdened by a misconception that safeguarding the human environment is somehow a minority activity. In part this is due to the invisible force of administrative and bureaucratic convenience which leads us to pigeon-hole problems and actions.

With this in mind I would like to outline the three priorities we must address before we can make any prevention-and-cure strategy really effective:

1.   First is the need for a cross-sectoral approach. Implementing an across-the-board strategy does not involve a major overhaul of the development process. In essence, it requires merely that the major players in the development process talk to each other before a project is given the go-ahead; we need the agronomists, the forestry experts, the economists, the hydrologists and so on to come out of their respective pigeon-holes to devise plans which will be of mutual benefit. Included in this process must be a comprehensive public relations exercise to find out precisely what the local people find acceptable. How often have we seen tree planting, reclamation and other development projects fail dismally to meet expectations because they have failed to enlist the enthusiasm and participation of those immediately affected?

2. Second is the need to build up local education and training capacities. Surely any strategy to roll back the wastelands is a futile exercise, unless it gives top priority to building up an indigenous ability to do the work of making sustainable development a success on the ground?
3. Third is the need to raise awareness of environmental issues among the general public. Of course that amorphous group, the decision makers, are our main target, but I have stated often enough my conviction that these people do not make their decisions in a vacuum. Unless the general public possesses an appreciation – no matter how vague, no matter how sketchy – of the vital importance of living resources conservation, precious little will be achieved.

The word 'microcosm', applied to any aspect of the problems India now faces, seems somehow rather inapt. Yet such is the scale of the global problem that in UNEP we must view India's advancing wasteland as but a microcosm of a world-wide process.

Land degradation

The areas most prone to soil degradation are the 30 million square kilometres of this planet's arid and semi-arid land. According to a study prepared in 1980 for the UN General Assembly, 80% of the world's total of irrigated land, rangeland and rain-fed cropland in arid and semi-arid areas are affected by desertification.

Each year some 6 million hectares are reduced to true desert-like conditions – one of the most complete forms of man-made wasteland. An additional 14 million hectares are reduced to zero or negative agricultural productivity. In India 61% of arable lands and at least 72% of non-agricultural lands are classified as degraded to a greater or lesser degree. In India and other dry, tropical regions of the world the situation is reminiscent of a fading carpet in which worn patches are being created and existing threadbare areas getting larger all the time. Our priorities then are to stop the threadbare areas growing in size and, simultaneously, to begin the work of repair.

At this point I think it would be helpful to be clear about what UNEP perceives to be the four main causes of farmland degradation. They are:

1. overcultivation;
2. deforestation;
3. overgrazing;
4. unskilled irrigation.

Each is made more acute by increasing human populations. The first three strip vegetation from the soil and deplete its organic and nutrient content, leaving it exposed to the eroding forces of the sun and wind. It can become as dry as dust, and blow away in the wind. World-wide we are losing, at a very conservative estimate, 25 000 million tonnes of precious topsoil each year.

The remaining subsoil can become hard and impervious. It can no longer absorb rain, and the water flows away over the surface, carrying away soil and cutting gullies which become deeper and wider year by year. In Tunisia it is calculated that almost 600 tons per square kilometre are lost each year from the central region bordering the Sahara. In India there are 3.7 million hectares of ravines and gullied land. The Indian National Commission on Agriculture calculates that 8000 hectares of land is degenerating into ravines annually.

Irrigation can do more harm than good if the water is not allowed to drain away from the soil. Poor drainage leaves the soil waterlogged. A high water table, and continual evaporation from the soil surface, can bring salt up from the subsoil and leave it in the topsoil. This 'salinization' process eventually makes the soil unusable. In my country, Egypt, some 30% of all irrigated land is affected. According to a recent Club du Sahel report, for each new hectare going under irrigation one hectare becomes too salty.

Drought and the lowered resilience of the land initiate a vicious circle of degradation. Human pressures build up as people vainly try to maintain crop and livestock production while soil fertility declines. In Rajasthan, only 20% of the state's arid land is considered by FAO to be suitable for rainfed cropping, yet such is the human pressure that the area under cultivation virtually doubled from 30% in 1951 to 60% in 1971, mainly at the expense of grazing lands and long fallows.

Desertification proper is confined to arid and semi-arid areas of the world. But it is part of a much larger picture of soil erosion in temperate regions, and in mountain watersheds such as the Himalayas. In Nepal, according to the World Bank, one quarter of the forest cover has gone in just 10 years.

Particularly at risk are island ecosystems with no hinterland to fall back on: a UN official who had recently visited the now inappropriately-named Cape Verde Islands off West Africa said the country looked 'like the moon from the air'. Haiti in the Caribbean has lost virtually all her forest which covered fully one half the country 30 years ago.

Nor are tropical rainforest areas like the Amazon Basin immune. Rainforest loss in recent years has caused drought in Indonesia, reduction in water levels supplying the Panama Canal, embryo deserts in northern Brazil, clogging of Thailand's traditional waterway transport system, and has drastically reduced the efficiency of HEP dams in the Philippines.

In all of these environments, man is taking more from the soil than he should. He is not putting back nutrients, not allowing the land time enough to regenerate its fertility under fallow and not restoring the vegetative cover that can protect the soil from erosion. To some extent, the severity of the problem has been masked by the intensive use of fertilizers. Where this is the case the problem will become apparent as more and more farmers find they are unable to pay for the increasingly expensive fertilizers made with fossil fuels.

These are but a sample of the causes and the multiple manifestations of the spreading global wasteland. As the Society for Promotion of Wastelands Development's publication and the Indian NGOs' recently published State of the Environment report reveal, these problems are repeated across the breadth of India. What is essential is not to allow ourselves to be daunted by these trends. If we allow that to happen, fatalism will set in and spread terminally like gangrene.

In UNEP we keep a watching brief and do our best to catalyze nations at the international and regional levels to take action. But nations are jealous of sovereignty and we must be careful not to trespass beyond the very proper limits laid down by the UN Charter. I repeat my message that it is for people in the ministries, farms, factories, local authorities and so on to undertake the work of pushing back the frontiers of the wasteland.

Surely it is time to affirm that we are in charge. With our technological and scientific capabilities, allied to political will, we can design a better future. Complex and daunting as the problems may seem, they are of our making and therefore are accessible to our solution.

In UNEP we have no patience with the argument that the money is not available to apply the World Conservation Strategy, the Plan of Action to Combat Desertification, the World Soils Programme, the clean water and sanitation strategies and so on. Even a small part of the US$1400 million the world spends each day on arms or the US$200 million on cigarettes would be enough to make the goal of sustainable development a reality. Arms spending buys only the illusion of security – the only real guarantee of lasting peace is sustainable and equitable economic growth.

So what are these solutions that lie within our reach? Here we return to the four main causes of land degradation.

Deforestation

Overshadowing all other causes of the world-wide destruction of forests are the demand for firewood, commercial timber exploitation and shifting cultivation. UNEP estimates that all types of tropical forest are disappearing at the rate of nearly 12 million hectares each year – lower than some estimates, but still a cause for alarm.

Shifting cultivation

Under the traditional system of shifting cultivation, the forested lands of the tropical regions provided an assured source of food, fodder and shelter for thousands of forest-farmers. In this century, increasing population pressure and spontaneous settlement have resulted in the current situation in which some 200 million people now live within forest areas and derive their livelihood from farming. These are the farmers, usually without title deed, who hack plots from the jungle and move on when the is soil is exhausted.

An indispensable part of stopping tropical rainforest destruction – and one too often ignored by conservationists – is land reform. In Latin America, particularly, new systems of land tenure are needed to give the landless a stake in their holdings; this will encourage them to make the sacrifices sometimes required (such as introducing a fallow cycle) to exploit the forest sustainably. Who can blame a farmer for not caring when he might be moved off his land at any moment?

Development assistance agencies should give top priority to starting, and continuing, support for projects which aim at restoring the balance between farmer and forest by, for example, helping him to plant nitrogen-fixing crops and trees with industrial uses, such as rubber.

In Malaysia an intensive land-use survey was carried out prior to settlement of some 9000 farming families in an area called the Jengka Triangle. Forest lands on steep slopes and river banks (comprising 60% of the area) were excluded from the settlement programme. Farmers were encouraged to settle on the better soil and to grow oil palm, a perennial tree crop that survives well in forest soils and performs a similar role to the natural forest in relation to soil protection and conservation of water.

Forestry is a long-term business (for example, 5–25 years are needed, on average, to produce significant volumes of fuelwood, poles and timber). Special incentives or subsidies are often necessary to induce farmers to take up tree planting. In the Philippines, for example, the incentive of a guaranteed high price for pulpwood, fuelwood and other forest products, coupled with the availability of credit, ample provision of tree seedlings, and the assurance of technical advice induced some 2500 farmers to participate in a successful tree-planting programme.

Commercial logging

In comparison with shifting cultivation, permanent deforestation resulting from commercial logging is not so significant. However, there are no clear-cut causal divisions. In areas lacking an adequate river network, commercial loggers drive roads through, and by so doing open up the forest for the shifting cultivators. In Asia nearly 75% of the permanent deforestation occurs in closed forest previously logged by timber companies.

Not only in India, but throughout the tropics the commercial foresters continue not so much to harvest the forest as to 'mine' it for high value species. Fewer than 20 out of the more than 300 species standing in the natural tropical forests account for more than 80% of total export trade. The economically accessible high-value species are being logged out at a faster rate than they are being replaced, and in the process the other species are being needlessly felled.

Reforestation with selected high-value species was running at fewer than 50 000 hectares a year in the tropical forest countries as a whole throughout the 1970s, and this needs to be increased five-fold if those countries – which currently depend on timber exports for foreign-exchange earnings – are to sustain the trade.

Governments should see to it that logging companies and their subsidiaries follow a strict code of conduct. The centrepiece of any such code of conduct should be controls ensuring that replanting is included in any new agreement. Currently under consideration are the institutional and financial arrangements for the proposed FAO/UNCTAD International Agreement on Tropical Timber, the first draft of which was presented to the UN by the Japanese Government last year. If consumer nations undertake to follow the Agreement's guidelines, we will have taken an important step towards rational tropical forest management.

The warm and wet climate of many tropical countries makes it possible to grow species of lowland pine on short rotations of 20–25 years, compared with the 50–80 years needed for coniferous trees in temperate countries such as Canada.

Compensatory plantations in eastern Brazil (4 million hectares) are now supplying about 50% of industrial wood needs for both domestic consumption and export. Many of these plantations have been established in areas that would otherwise have been alternately baked by the tropical sun and washed away by tropical rainstorms into a wasteland beyond hope of recovery.

Firewood

The third main cause of forest degradation is meeting the demand for firewood and charcoal. About 1.5 billion people in the developing world depend on wood for their domestic energy requirements. In Africa south of the Sahara, 60–65% of all energy needs are still met by firewood and charcoal. Annual removal for fuelwood is reckoned to be more than 700 million cubic metres in Asia and more than 200 million cubic metres in Tropical America, and there is no prospect in the foreseeable future of a slackening in demand.

In central Tanzania and the uplands of Nepal, recent studies indicate that it now requires 250–300 man-days of work to provide the annual fuelwood requirements of a household of five persons. (In fact, 'woman-days' would be more accurate since it is they who must collect the firewood.) Fuelwood shortages in the vicinity of urban townships such as Niamey, the capital of Niger, have led to the gradual destruction of all savannah woodlands within a 50-kilometre radius of the capital.

India's Planning Commission has estimated that, even if all the current and projected tree-planting plans are implemented, these will help to produce only about 49 million tonnes of fuelwood each year on a sustained basis, as against a requirement estimated at 133 million tonnes of fuelwood annually. India's problems are a mirror for the Third World's as a whole. To meet a demand of this scale will require increased rates of reforestation in the developing countries on a scale more than five times that of current annual levels. In some smaller African countries, annual planting rates would need to be increased up to 15 times over current rates.

Reaching these tree-planting targets means that the authorities not only have to chart plans and mobilize necessary human and financial resources, but also have to make the right decisions about which trees to plant: fast growing trees are required, but to be really useful, trees should at the same time be nitrogen-fixers, have industrial uses, provide animal fodder and food, and be good for making charcoal.

We must call in the wastelands to meet this demand. And I do not just mean wasteland in the conventional sense of previously logged areas or abandoned mining sites, but also land that is wasted because it is not being utilized: for example, village commons, roadsides, the margins of the fields, and railway and canal sidings. In PR China and South Korea – two nations with very different political and economic systems – a balance between supply and demand is reported to have been struck. It can be done.

This is the supply side, but what about demand? Nations must make some very tough decisions about, for example, which renewable energy technologies to invest in. To eke out firewood supplies, fuel-efficient stoves are one obvious answer. The need here is to make decisions not merely on the basis of efficiency, but also on what is socially acceptable. For example, in some Nepalese villages open fires are still preferred to the new stoves because smoke keeps down insect infestation in the roof thatching. In this case the solution might be stoves and corrugated iron roofing.

With tree-planting drives, and with the promotion of fuel-efficient stoves, solar cookers and other environmentally-sound hardwares, it has far too commonly been the practice of outsiders to come in and tell locals what is good for them. They resent it and it does not work. Extension workers should take more account of indigenous needs and traditions. Whatever the solution, popular participation is one element essential to the success of any wasteland utilization programme. I cannot stress too often that outside financial support for projects is a futile exercise unless they have the potential to secure the enthusiasm and participation of local people.

Overcultivation

Degraded land is land *en route* to a wasteland. In effect it is wounded land – but we know it can be nursed back to health and full productivity. Our dilemma, and our challenge, is how to do this and at the same time not merely maintain agricultural productivity, but increase it to feed the near two billion extra people projected to be with us in the year 2000. Were it not for our innovative scientific and technological capabilities we would, indeed, be in a hopeless position.

First and foremost we need to direct much more effort into producing improved varieties of conventional crops, such as sorghum and millet. But they would need to be as resistant to drought as traditional varieties, so as not to cause a catastrophe for farmers in dry years.

Farmers will be unwilling to commit themselves to investment in intensified cropping without firm assurances from the government, not only on credit, but also on the prospect of attractive prices for their crops.

Greater productivity will be meaningless unless there are good facilities in the area for storing grain as a reserve for drought years when crops fail. Mauritania has just completed a national network of grain silos in a $10.5 million programme, co-financed by the African Development Bank and the Netherlands Government through the United Nations.

Higher yields can enable farmers to break out of the vicious circle of reducing fallow periods and constantly expanding the area under cultivation in order to maintain production at previous levels. It could therefore be possible to satisfy increased demand for food, at the same time as reverting to traditional rotation systems, such as the four-year sequence of fallow, millet, cowpeas and groundnuts that is still common in parts of the Sahel.

Yields can also be increased by improved methods of cultivation, such as deep tillage. Often farmers do not turn under organic matter because they are reluctant to incorporate potential fodder material into the soil. This could be overcome by the growing of trees such as Gliricidia, which produce plentiful foliage that can be used for both fodder and 'green manure'.

Villagers' nutrition and overall food production could be improved by establishing irrigated vegetable gardens or fruit orchards near wells. An experimental project in Mauritania is investigating this approach by promoting the creation of small market gardens in the vicinity of 36 deep wells drilled as part of a joint UN/African Development Bank project.

Greater diversity of crops could also be obtained by trials of underexploited drought-resistant species, such as those recommended by the US National Academy of Sciences: grain crops like grain amaranth and Pima-Papago 60-day maize, and nitrogen-fixing beans like the mottled lima bean and vegetables like Hopi and Papago Cushaw squash. Whether such improvements in rainfed cropping do take place depends as much on the will of governments as on the courage and imagination of farmers.

Bad irrigation

Is irrigation a dead end or a short cut to increased food production? Certainly irrigation can lead to six-fold increases in cereal yields and up to five-fold increases in root crops. At present irrigation accounts for some 13% of all land under cultivation. In almost every arid and semi-arid country, large-scale plans are underway to increase that area dramatically. However, mostly through a combination of bad planning and poor maintenance, half a million hectares of irrigated land throughout the arid zones becomes desertified each year.

Typical in many ways is Iraq's Greater Mussayeb Irrigation Project started in 1953. At the outset, 60% of the land was saline; irrigated land was poorly surveyed and levelled; there were no field trials to determine the water requirements of different crops; managerial staff was inadequate; irrigation canals silted up; and drains were badly maintained. By 1969 waterlogging was common, and two thirds of the soil was saline. Yields fell, and barley was substituted widely for wheat. Many farmers reverted to the traditional 'nirin' system – cultivating cereals under flood irrigation until the land becomes saline, then leaving it fallow so that the rain can leach salts out of the soil.

In Pakistan virtually half the soil is affected by salinity; in Soviet Central Asia, despite great efforts aimed at improving drainage, in the 1970s 86% of the more than one quarter of a million hectares which had been irrigated was saline or waterlogged. In Western Australia 170000 hectares of irrigated land is out of production.

The irrigation potential in the tropics is huge – 4 million hectares in the major river basin areas of the Sahel alone. But have we learnt from our mistakes? Can we refuse the temptation to be dazzled by the prospect of short-lived bumper increases in yield? There are hopeful signs. USAID, for example, has changed its policy and will now concentrate on small- to medium-scale schemes that include large elements of farmer participation. However, in line with the 1977 UN Conference on Desertification's recommendation, I believe the first priority should be to rehabilitate the irrigated wastelands.

Overgrazing

The loss of balance between people, resources, environment and development is nowhere more evident than in the damage caused by overstocking. Two main factors have reduced the amount of pasture available to herds, forcing them onto increasingly marginal lands in search of food:

- first has been the expansion of rainfed cropping onto marginal lands normally used for grazing;
- second has been the increase in the cultivation of cash crops and ranching on lands traditionally grazed by herds of the nomad and semi-nomad pastoralists.

Being restricted to a rump of their former lands has not stopped the pastoralists trying to increase herd sizes to feed increasing populations and to improve the standing of individuals in their community – a community in which wealth and status are viewed almost exclusively in terms of how many livestock one owns. It only needs a drought, the occurrence of which is inevitable, to touch off a crisis.

There is a third factor. The authorities in the modern nation-state tend not to like nomads and those who live a semi-nomadic existence. They are difficult to tax, they tend to pay scant attention to political boundaries and, above all, their way of life seems anachronistic. The response nearly everywhere has been to try and settle them through the provision of boreholes, churches and shops. The result has been that around the new settlements some of this planet's most complete wastelands are now to be found.

However, a trickle of evidence is now emerging that governments are beginning to realize that an amended form of nomadism is the only practical way to exploit arid pasturelands sustainably. In Saudi Arabia, for example, efforts are being made to preserve the traditional Bedu practices. In Northern Kenya a UNESCO team of researchers has produced a management plan which, if implemented, would retain nomadic practices but combine them with modern marketing, veterinary services, destocking programmes and a system of reserves to be used only in times of drought-induced emergencies.

In essence, I believe there are six principal ways by which livestock productivity could be increased, without damaging the environment. However, it should be said that there is not much consensus among the experts as to the relative value of these steps. They are:

- improving the quality of animals by disease control, breeding, and stock reduction;

- improving pastures by reseeding, allowing time for regeneration, planting new fodder/forage crops, and other measures;
- increasing offtake, that is the number of animals sold each year for slaughter;
- improving infrastructure by spacing wells in accordance with available water resources, building all-weather roads and making other improvements along routes to market, and establishing feedlots and rural-based abatoirs;
- changing the organization of pastoralism by taking measures to restrict or regulate grazing in certain areas, and developing regional livestock raising schemes;
- basing all new livestock development plans on fine calculations of carrying capacity

Wasteland created by industrial pollution

Pollution of the air and water can be considered as the presence of a resource in the wrong place at the wrong time. India's National Environment Engineering and Research Institute estimates that 70% of the available water in India is polluted. Air pollution is helping to turn the cities of the Third World into a form of wasteland. India's Sixth Five Year Plan states that the levels of SO_2 and particulate matter in major cities such as Calcutta, Delhi and Bombay exceed the permissible limits set by the World Health Organization.

One of the aspects of global environment degradation which is most worrying is the pollution in the newly industrializing nations such as Brazil and South Korea. Mexico City and Ankara, for example, have experienced some of the worst air pollution episodes ever recorded – and that is not discounting the notorious London smogs of the late 1940s.

It is a '*déjà-vu*' problem. The developed nations, sometimes at enormous expense, have partly begun to control their industrial pollution. However, the opportunity not to repeat their mistakes for the most part has not been siezed. With so few resources to hand in the developing countries to control emissions, the situation would be gloomy indeed, were it not for the fact that we are now able to demonstrate, in terms of boosted profits and output, that preventing pollution and recycling waste pays.

UNEP, the OECD and other organizations have been subjecting the environment to cost–benefit analyses. Though the state of the art leaves considerable room for improvement, we have been able to demonstrate that nations with environmental programmes in place can make net savings in GNP, that jobs are created, that control measures have been an insignificant cause of inflation, and so on.

Thus, if pressed, we can now couch the benefits of environmental improvement in hard cash terms. For example, in 1979 the US Council on Environmental Quality showed that the benefits of measured improvement in air quality since 1970 could be valued at US$21.4 billion per year. By 1985, in the USA benefits from water pollution control could be worth US$12.3 billion annually.

Most encouraging of all has been industry's growing awareness that profit margins can be increased through waste recycling and the voluntary introduction of resource-saving technologies. For example, new processes for sulphuric acid manufacture have reduced emissions from 17.5 to 3.5 kg/ton of sulphuric acid; in the pulp and paper industry, recently constructed mills have reduced liquid effluent discharge from 180 to 70m³/ton of pulp. Other industries reduced their consumption of water and improved water recycling. In some cases profits were increased by as much as 40% as a result.

In some countries industry has shown commendable initiative in switching and then utilizing wastes. For example, the Japanese cement industry is currently using 35 million used car tyres per year in the cement-making process. By burning those tyres, two million kilolitres of imported crude oil are being saved every year. Many large companies have now added divisions to provide environmental goods and services. They include Boeing, Exxon, Dow Chemicals, Shell, BP, Krupp, Philips and ICI.

The cost–benefit approach has demonstrated that the strategies which bring the greatest economic return are the preventive ones. The 3M Corporation was one of the pioneers. Its 'Pollution Prevention Pays' PPP programme, based on conservation-oriented technology, has been a great success story: between 1975 and 1981, the PPP programme produced total savings of US$97 million. To make an equivalent amount in earnings the company would require US$300 million of additional sales per year.

Thus the new direction for the next decade is the development of recycling and preventive technologies. And perhaps most encouraging, UNEP has logged instances where industrial performance has been dramatically enhanced by the introduction of conservation-oriented technologies in countries as different as Belize, PR China, South Korea and Brazil. The rapidly industrializing countries should beware the salesman from the developed countries who want to sell them outdated plant which, though initially cheap, will prove much more costly in the long run.

Although increasingly numerous, the instances of introduction of non-waste and low-waste technologies are still the exceptions which have not yet proved the rule. It is for governments, through incentives, and industry, at its own initiative, to step up over the next decade the production and in-place application of such technologies.

Perception of wastelands and candidates for rehabilitation

The areas devastated by mining, those turned into true, desert-like land, deforested mountain slopes reduced to bare rock, even the areas subjected to nuclear test explosions or chemical warfare, are the complete wastelands – if you like, our environmental bomb sites. There is no doubt that, with a great expenditure of effort and resources, we can reclaim those areas.

It is necessary to make a series of very difficult decisions about the best candidates for rehabilitation. There has been mention of the need to apply the first world war concept of 'triage' to conservation. During that war the wounded were put in three categories, with the minimum of medical effort expended on those most likely to die. Most effort was thus spared for those who were most likely to recover to fight again – a brutal but practical set of priorities.

I would suggest that in rehabilitating wasteland areas we need to make similarly harsh decisions over where to concentrate our limited resources. If we try to do everything, we may end up achieving nothing. We need to concentrate on the partially degraded lands, the lands most likely to make a full recovery, and on the areas where the local authorities and the local populace show most willingness to participate in the endeavour. The seriousness of the situation leaves no other option. We may lose some battles, but such a strategy gives us hope of winning a war through conditional surrender.

We also need to improve our definition of the concept of a wasted area. This too will help us to focus action. Roadside verges, railway cuttings and so on are such wasted areas. In the Philippines they are planting Leucaena trees in graveyards.

Other categories of wasted land are many and diverse. In Africa, for instance, vast areas of good farmland are wasted because we have not put sufficient effort into eradicating the vectors of river blindness and sleeping sickness. And what of the millions of square kilometres of potentially productive farmland reserved for the exclusive use of the military? And what of the huge areas of Central America's tropical forest devastated for cattle ranches which provide the USA's hamburger chains with just 2% of their beef requirement? So too is prime agricultural land needlessly going under tarmac, urban sprawl and new factories. With the land speculation lobby controlled and the implementation of sound planning, most new developments could take place on the least productive sites – perhaps even on the 'mortally wounded' of our wasteland triage.

In conclusion, we have exhausted some of our options. But a great deal more, in various states of productivity, still remain to us. With sound management we can restore many of our wounded areas to full health. The India Society for Promotion of Wastelands Development has a vitally important role to play in rolling back India's wasteland – and I wish it every success in its endeavours.

8

Combining the best of the old with the best of the new

Statement to the First Indian National Environment Congress

New Delhi, India, December 1982

Nineteen eighty-two has been a year of stock-taking for the environmental movement, for it marks the tenth anniversary of the 1972 Stockholm Conference on the Human Environment, a Conference that not only created the UN Environment Programme but which gave direction, form and expression to the international environment movement.

With this tenth anniversary year drawing to a close, we are about to embark on what, for us in UNEP, is the beginning of a new and critical decade. Inevitably, as we do so our concern for what has happened over the previous 10 years recedes as we set our sights on the objectives governments agreed upon at the commemorative conference held in Nairobi in May.

Nevertheless, I think it would be appropriate to start my address by recalling the two main objectives the Stockholm Conference set for the global community. They were:

- to increase our knowledge of the environment;
- to safeguard and enhance the environment for present and future generations.

I reported to the 105 governments attending the Nairobi Conference that, although we can make a strong case for claiming fulfillment of the former objective, we are even further away now than in 1972 from enhancing our human environment.

Destruction of the environment

In 1982 the global community still tolerates a process of global development which destroys more than it develops. Lamentably, there are statistics in plenty to support this assertion.

The tropical forests are disappearing at a rate close to 12 million hectares per year. Those forests are the great genetic powerhouses. At a conservative estimate, one in 10 flowering plant species – some of which could be of incalculable benefit to agriculture, medicine and industry – are threatened with extinction. The insatiable demand for firewood and charcoal is perhaps the single most environmentally destructive force we have to deal with. In Africa south of the Sahara, for instance, two thirds of all energy demand is still met by firewood. Despite progress in renewable energy development, we need to be planting trees at five times the present rate to meet current cooking and other firewood demands.

Today 450 million people are chronically undernourished. If we carry on as we are, there could be double that number by the year 2000. Meanwhile 20 million

45

hectares of the farming land so desperately needed to feed those alive today are degraded each year to zero or negative productivity. World-wide, an estimated 70 000 square kilometres of prime agricultural land was lost in the 1970s to encroaching cities. On top of that, through avoidable post-harvest grain losses, each year we waste at least the equivalent of 10% of world cereal production.

The slums and shanty towns of the developing world are growing at the rate of 8% each year. Nearly half the people living in Third World cities have no access to sanitary facilities. In the rural areas the situation is even worse, with 87% deprived of proper sanitation. Current trends indicate that by the turn of the century nearly a billion people in the cities will be denied access to shelter, clean water and the other basic human needs.

Corrective strategies

Even with the projected 2 billion people to be added to the world population by the year 2000, this one world could still be enough to satisfy our needs. Over the previous decade, through analysis of the problems we have come up with a series of corrective strategies, such as the Brandt Report, the World Conservation Strategy, Action Plans to combat desertification and to provide clean water, and now a soil conservation plan and a programme for rational use of tropical forests which, if implemented, would reverse destructive development.

If we fail to change course over the next 10 years our sorry legacy to the next generation will be an empty food cupboard, and worse, no means of replenishing it. This wasting of resources has already begun to destabilize the world economy and pose a tangible threat to the security of nations. Yet governments continue with wasteful and destructive development policies.

The question is why? Why do so many nations persist with these self-defeating policies? Many reasons have been advanced. They include timidity or indifference among decision makers, lack of political will, economic conservatism, and entrenched business interests. These are obviously part of the answer. However, I believe the root cause of our environmental crisis is that governments and people still take this planet's threatened living and other renewable resources for granted.

Even if we discount aesthetic, health and similar non-economic considerations, we find, in terms of strict economic accounting, that fresh water, watersheds, soil, forests and so on are every bit as valuable as, say, industrial plant, fossil fuels and other mineral deposits, but, inexplicably, we do not count them so.

Consider soil. It can take nature hundreds of years to accumulate top soil to a hoe's depth. If stripped of protective vegetation or ploughed too deeply, it can be removed in months. The result is loss of agricultural productivity, silting of river systems and, sometimes, tragic loss of life in the wake of floods.

According to a recent estimate, the small African country of Lesotho has lost 100 000 hectares of arable land in just four years; a disaster for a poor nation where cultivable land is at a premium. No wonder that in Lesotho they have a grim saying that 'their only exports are "soil and cheap labour" '.

The story is the same, or similar, in the rest of Africa where the land is failing to sustain a rapidly increasing population. In Asia, and also in Latin America, food production has kept pace with the expanding population, but only just.

So far as I know, there are no calculations on a world-wide scale of the cost of the 25 000 million tonnes or so of soil washed into the seas each year. But is has been estimated that the current annual loss of production due to desertification is in the order of US$26 billion. By contrast, it would cost only US$2.4 billion each year to carry out the world plan to combat desertification.

The overlapping problems of soil erosion and desertification have a common cause in mismanagement, which all too often takes the form of no management at all. Nor is it a problem for the developing nations alone; in the USA it is estimated that more soil was lost during the 1970s than during the great dustbowl episodes of the 1930s. Again, much of the problem, I believe, stems from this indifference to renewable resources – we continue not so much to use soil as to mine it. The same can be said for our forests or fish stocks.

At the Nairobi Conference in May, soil loss was rightly described as a haemorrhage, weakening and draining away the life of this planet. Water contamination, destruction of drainage systems, air pollution, deforestation, expenditure on arms and post-harvest losses are other haemorrhages which we can, and must, stop.

Maintenance of the environmental infrastructure

Like the man-made environment, our natural environment – the biosphere – is a complex infrastructure and we need to treat it as such. Like a nation's social and economic infrastructure, it must be managed and properly maintained.

No country would deliberately neglect to maintain the component parts of its socioeconomic infrastructure, such as the water supply and sanitation system, transport network or national electricity grid, because a collapse of one part results in a partial or total breakdown of the whole infrastructure. It is the same with the natural infrastructure, where a parallel mechanism of cause and effect applies. Thus, a depletion of the ozone layer in the atmosphere will allow increased quantities of ultra violet light to filter through to the earth's surface where it may damage human, plant and animal life; wanton tree felling coupled with increased carbon dioxide emission from fossil fuel sources can alter precipitation patterns with serious consequences for farming; and unsustainable withdrawals of underground water supplies can cause salinization and land subsidence.

Painfully, and far too slowly, we are realizing that in many sectors we have already exceeded the limits. Even so, based on recent experience, I believe nations will persist with self-defeating policies even to the point of disaster unless, and until, they apply a strict system of environmental accounting.

The gathering pace at which our environmental infrastructure is breaking down is one of the fundamental but largely unrecognized causes of the world economic depression which refuses to go away. The way out of the global slump lies in pursuing economic policies aimed at achieving sustainable and equitable growth.

Management of the environment

In Nairobi in May, 105 nations made a formal declaration of commitment to what amounts to a global plan of action to maintain and repair our environment infrastructure. At the Nairobi meeting there was little talk of 'environment protection'. Instead the emphasis was on sustainable development through competent management of resources. The consensus at Nairobi was that we need to revolutionize attitudes among people and governments, with the aim of changing wasteful life-styles and consumption patterns. We need to use resources on the basis of careful calculations of limits for continuous yield. We need to make decisions which take into consideration the carrying capacity of ecosystems.

The governments represented in Nairobi recognized that sound management, in turn, depends on building up local management capabilities through education and training. They also came to the conclusion that action to solve the environmental crisis will not be effective unless the general public is given a much clearer idea of the issues and what is at stake for their future livelihood. They called for priority action to be directed to awakening public concern through the NGOs and the mass communication media.

What also emerged at Nairobi was recognition of the importance of more research and data collection. There is an obvious need to clear up the many environment imponderables. No action on problems like the potentially grave effects of the build-up of carbon dioxide in the atmosphere can be fully effective until scientists reach a consensus on the precise nature of this problem, and on others which are still unresolved or the subject of dispute.

UNEP – the world's only global environment intergovernmental organization – came out of that conference with its catalytic brief reinforced. But at a time when we are being overwhelmed by requests for guidance and assistance from the developing world, the pledges of increased contributions have not allowed for a growth – in real terms – of our Environment Fund. In UNEP we see an absurd discrepancy between the serious concern registered by governments, from North and South, about the deteriorating state of the environment and the means put at our disposal to tackle the problems identified.

Implicit in the Nairobi Declaration is an awareness that the spread of human activity has caused this planet to shrink to a global neighbourhood. The various sections of the world community have become so close and interdependent that errors and omissions in one country, or one region, have global implications. What nations do collectively at the international and regional levels is thus an indispensible part of solving our shared problems. But that said, in the last analysis, our common salvation rests on what countries – governments and people – accomplish within their own borders.

Meetings such as this Environment Congress are important for determining new national orientations and for reaching agreement on what has to be done. But the real work of managing our environment rationally depends on the quality of the day to day decisions made in the ministries, provincial and local authorities, NGOs, businesses, villages and farms the world over.

India has a special role to play. As an acknowledged leader of the non-aligned movement, and as a country which has consistently pursued the self-reliant and independent course of economic development set by Prime Minister Nehru, India has much to teach its fellow Third World travellers.

The Nairobi Conference laid much stress on the need to step up TC–DC (technical cooperation among developing countries) in environmental management. I believe that through the medium of the UN, and other relevant bodies, developing nations could benefit much more than they do now from India's pioneering work in pharmaceuticals, in irrigation, in birth control, in biogas and solar energy development and in other areas within the compass of the environment programme. In India itself Gujarat's social forestry policies, Rajasthan's antidesertification programme, and Uttar Pradesh's Chipko movement are examples others would do well to adapt and apply.

At Stockholm, and now more than ever, we in the environment movement hold firmly to the view that of all the resources available to all nations, people are the most precious. I have an unshakeable belief that through human enterprise, inventiveness, application and sheer common sense, we will arrest the process of resource destruction. But in our belief that a restoration of the infrastructure lies in the sum of individual action, we must guard against a new danger, one that is very much a product of the 1970s – fatalism.

In fatalism – the attitude that no matter how hard we try, things will not improve so let us not bother – resides our greatest collective peril. The oil price hikes, the worsening terms of trade for the developing nations, and the failure to provide basic human needs have sapped confidence to a dangerous extent. Notwithstanding this danger, I take profound encouragement from signs, emanating from all countries and at all levels, of a developing commitment to tackle the base causes of our problems. We need look no further than India's newly established Society for Wastelands Development for such a sign. By taking as its guiding principle, the notion that there need be no such thing as a wasteland or waste of any kind, this new Society, like literally thousands of citizens action groups around the world, will counter fatalism and in so doing help roll back an advancing wasteland.

At the Stockholm Conference the Prime Minister of India, Mrs Indira Ghandi, told delegates that the worst form of pollution was poverty. At that time, to all but the most enlightened at the Conference, it was a revolutionary idea. But in the intervening years the wisdom of these words has been graphically demonstrated as underdevelopment has forced the underpriveleged to exhaust the resources on which their future survival, let alone prosperity, depends.

What was also recognized at Stockholm, and reemphasized at Nairobi, is the equally insidious threat from the opposite end of the economic spectrum – wastage of resources through mishandled wealth. The developing world's renewable resources are their environmental capital. The demands of a rich minority of nations are forcing the poorer nations to squander that capital. From within the Third World the privileged nations, or groups within nations, add to that burden by copying Western consumption patterns.

The Third World is the source of one third of all the primary products, excluding fuel, imported by the OECD nations. But the prices paid for those products have not kept pace with the spiralling costs of their manufactured goods. The conditions of trade were actually better for the developing nations 20 years ago than they are today.

The need for sustainable patterns of development

At a conceptual level there is now a broad consensus that the industrialized nations cannot remain immune from an environmentally-induced economic collapse in the developing world. The need for a more just, less consumptive economic order has won wide acceptance but we have yet to see these concepts applied in any meaningful way. What is so depressing is that long before the New Economic Order was tabled on the international agenda, the need for alternative, more sustainable patterns of development was recognized. At the turn of the century, the great Mahatma Gandhi expressed this need extremely well when someone asked if he would like a free India to become like Great Britain? 'Certainly not', he replied, 'if it took Britain half the resources of the globe to be what it is today, how many globes would India need?' On another score, Mahatma Gandhi was a powerful voice for the need to maintain the old traditions. He believed that the past has much to teach the present – a message which has lost none of its relevance.

For thousands of years, in some cases, villagers and townspeople have used local resources prudently so that they have been handed over intact from one generation to the next; often if has been accomplished unconsciously as they observed religious, cultural and other practices finely honed for their environment. In terms of material possessions, these people were poor, but in their respect for what nowadays we refer to as the balance between people, resources, environment and development they were rich indeed.

In 1982, the global imperative is a restoration of that balance. This does not mean I am calling for a brake on modern development through a return to a pre-industrial economy. Rather we must apply alternative and less wasteful development patterns which will revive, and where it has already disappeared without trace, resurrect a respect for living resources. Simultaneously, governments must do all they can to use the material benefits of the most advanced technological developments to increase the standard of living for all their people. In short, we must combine the best of the old with the best of the new.

On the occasion of the launch of the World Conservation Strategy the Indian Prime Minister, Mrs Gandhi, made an appeal for man to retain a communion with the earth, once again to 'put his ear to the ground so that the earth can whisper its secrets to him'. At the same time, she reiterated the need to achieve growth in the economic and social well-being of people throughout the world by following a course of ecologically sound development. In effect, she was also calling for a policy of combining the best of the old with the best of the new.

By steadfastly observing this policy we will, albeit belatedly, begin in earnest the work of fulfilling the Stockholm objective of enhancing the environment for this generation and for those yet to come.

9

A duty to hope – new directions for science and technology

Statement to the Barbara Ward/Rene Dubos Memorial Seminar

New York, USA, February 1983

Rene Dubos and Barbara Ward were, above all, great reconcilers. At Stockholm and beyond they worked to reconcile the industrialized and the developing nations, and they were among the first to articulate the complementarity of environment management and economic growth. This convocation is a tribute to their continuing influence. We are here at this seminar to exchange views and opinions and, hopefully, in so doing to continue their work of reconciliation.

To the end of their lives, Rene Dubos and Barbara Ward shared a conviction that this one world is enough to provide for all of us. I share their conviction. But let us be clear; if we are to realize the goal set by governments at the Nairobi commemorative conference in May 1982 to make maximum sustainable and fair use of this planet's resources, we must revolutionize our attitude to science and technology and the values we attach to their usage.

In reviewing the progress made in science and technology since Stockholm we can say there have been many positive developments for the environment. But that progress has served to show only what might have been achieved and, more important, what could be achieved in the second decade after Stockholm.

We have begun to question seriously whether science and technology must always be employed in the service of making things bigger and faster. We have a much keener appreciation now than a decade ago of the limits of this planet's resources, and we have begun, in a few sectors, to use science and technology in the service of sustainable development.

What lies within our grasp, today, is the power to use science to shape a better future. Many technologies are far less harmful to our environment than they were 20, even 10 years ago. We can also calculate the long-range consequences of technology. But we still display insufficient will and little confidence in exercising that controlling power, and we have failed dismally in the task of setting worthwhile social goals.

Rene Dubos observed that our relationship to science and technology need not be passive. Dubos considered that we have had a tendency to surrender too often to technological imperatives instead of striving for other desirable human values. It is not a question of sitting back and seeing where technology is taking us, it is a matter, he said, of using science and technology to take us where we want to go.

It is time to affirm that we are in charge. Complex and daunting as the problems may seem, they are of our making and, therefore, are accessible to our solution. As Barbara Ward said: 'No problem is insoluble in the creation of a balanced and conserving planet save humanity itself'. What I hope to indicate in this statement is that these daunting environmental problems are easily within reach of current technology to solve.

Our chief environmental insight

We have gone a long way to meeting the Stockholm objective of increasing our knowledge of the environment. In 1983 we know immeasurably more about the interrelatedness of our environment. For example, we know that the global cycles of carbon, oxygen, nitrogen and sulphur are inextricably linked. The notion that 'everything is connected to everything else' has been dismissed as a commonplace. Significantly, Barbara Ward recognized that 'commonplace' to be 'our chief environmental insight'.

The developing knowledge of the interconnectedness of our environment surfaced perhaps most poignantly in the World Conservation Strategy, and the Plan of Action to Combat Desertification and the World Soils Policy which base their appeal on the need for a cross-sectoral approach. Mostly through harsh experience, we have found that positive action in one sector can be frustrated by neglect in another. Thus the useful life of some HEP dams has been halved because a sound watershed policy has not been applied elsewhere; or otherwise effective pollution controls have been frustrated because nations situated upwind or upstream have not taken reciprocal action.

The obvious implication is a need for complementary technologies. Using shared natural resources equitably and sustainably requires unrestricted access to those technologies. Share and share alike might be another commonplace, but it serves the benign self-interest of all nations to apply the principle.

Filling in the blanks

If science has provided us with a much better picture of the dynamic workings of our environment, it has also exposed the blanks – the areas where there is not sufficient scientific knowledge to allow us to reach a consensus on how great the threat is, still less on what remedial action is needed. We now urgently need to fill those blanks, to clear up what I call the environmental imponderables.

Depletion of the ozone layer, the build-up of carbon dioxide and acid rain are the major atmospheric imponderables we have to clear up. What, for example, is the exact impact of ozone depletion? We cannot say for sure what effect the continuing build-up of carbon dioxide in the atmosphere will have on the climate. Will it turn the earth into some kind of Venusian hothouse? Or will the greenhouse effect be offset by the shielding effect of more industrial particles? Who will be the 'winners' and 'losers' in such a process? What will be the likely political ramifications? Acid rain has recently been described as a kind of chemical warfare. But the atmospheric processes which create acid precipitation are not fully understood, and there is little consesus as to which nations and individual industrial sectors are the main perpetrators of this 'chemical warfare'.

More research is needed before a consensus can be reached on these and other imponderables. But this should not be used as an excuse for delaying action. If we wait until we are absolutely sure, it could be too late.

We are scarcely better informed about the marine environment. Our knowledge of marine ecosystems is at best sketchy. Until we find out more about how marine systems function how can we, for example, fix safe fish catches?

The fish harvest provides the world with 6% of all the protein consumed: in 32 nations fish account for more than a third of the total protein intake. From 1900 the marine harvest increased steadily until the 1970s when it began to level off. Expanding the limits requires management and international cooperation, but such developments will amount to little unless much more effort is put into establishing maximum sustainable yield levels. Top priority in scientific research should go to the tropical seas about which we know little, and to the Arctic and the Southern Ocean, about which we know even less.

We need also to discover much more concerning the long-term effects of oil spills, of pesticides and other forms of marine pollution. Just as important, we need to find out a great deal more about the value of wetlands and coastal shallows which provide not only the main fish and crustacean breeding and nursery grounds but also help regulate drainage and provide a barrier to coastal erosion. These are nations' liquid assets. Unless, and until, marine science provides the decision maker and public with evidence of the economic value of these liquid assets they will continue to be haphazardly drained and otherwise destroyed.

Even on land there are blanks. Take genetic resources. Just 1% of the flowering plant species have been tested for their potential usefulness to mankind. Yet we must contemplate the disappearance between now and the end of the century of between 10% and 20% of the current range of wild plant and animal species. Genetic engineering is a rapidly advancing science. But the scientists must have the basic raw material to work with. Who knows? May be the unnoticed disappearance of, say, a fungus could be robbing us of a penicillin-like medical breakthrough?

Our misdirections

Science and technology can and should be the friends of the environment. Notwithstanding the positive developments (which I shall amplify later), we have to say that technology is still used most often in the service of a development process which destroys more than it develops.

The prime example, of course, is the arms industry. In spending US$1400 million on what is euphemistically called 'defence' each day, the world spends 20 times more on weapons than it does on development assistance. Some 40% of all R&D goes to military ends. It is, I confess, a mystery to me why scientists, in political systems that allow free choice, allow themselves to waste their talents on destruction. If they are duped, it is because they allow themselves to be so.

A closely connected and equally serious misdirection is the effort presently being ploughed into advancing technologies that are of exclusive benefit to the first class passengers on our global aeroplane. Thus we read that the silicon chip will soon be out of date, that the barrier to improving computer performance is now the speed of light and so on. One wonders how we can reconcile the concentration on this technical wizardry with the neglect of technologies which will improve the conditions of life for the second class passengers?

I have a suspicion that we are being mesmerized by the technology itself; that the preoccupation with perfecting the technology is creating its own impetus. We have to ask how often are potential social benefits taken into account before the go-ahead is given to new technology development.

Rene Dubos in *Celebrations of Life* observed that 'Few indeed are the new products and processes of modern technology which are introduced to meet fundamental needs of humankind. Most of them appeal instead to the desire for change for change's sake . . .'.

The rapid advances in satellite technologies are, I believe, a case in point. From outer space we now have the ability to peer into all our global nooks and crannies. Granted, we are using the currently workable earth resources, space technology and the microprocessing spin-offs to improve our understanding of how the biosphere functions and to improve the decision-making process with respect to food production and use of natural resources. But these are the exceptions; not nearly enough effort is going into making sure satellite technology serves our practical and peaceful needs. At the end of the 1970s, one military satellite was being launched every three days.

We now need a shift of emphasis away from further perfection of techniques by the suppliers towards peaceful, practical, in-place applications to meet the needs of users. There is a waryness in the developing world of the speed with which the West has embraced the idea that so-called 'appropriate technology' is best for them. This waryness emerged clearly at the UN Conference on New and Renewable Sources of Energy. The Latin American nations, in particular, were disturbed by what they saw as an overemphasis on improved jikos and tree planting. They appreciate their importance but also want advanced technologies, as part of an overall programme of mixed technologies, which will help the developing nations climb the steps of the technological ladder.

What is abundantly clear is that all the rich world's technological wizardry will not be able to put the living resources of the planet back together again without major reorientations in direction. There are reasons of self-interest to make this change: in the tourist section of our global aeroplane there is overloading and growing dissatisfaction that can only lead to chaos, and the mayhem will spread to the first class compartment.

New directions

The challenge over the next decade will be to build on the positive developments to create a new regime of science and technology that will use the resources of this planet prudently. In this, our aim should be to hand those resources not only intact, but enhanced, to the next generation.

Industry

The 1950s and 1960s were an era of rapid industrial development, with little regard for environmental consequences. The origins of the environment movement in the West can be traced to a groundswell revulsion against these practices.

From a safe distance of some 10–15 years we can now say there was something of an overreaction. A new species, 'the environmentalist', appeared, saying some intemperate things about the value of economic growth *per se*, and pressing for costly pollution controls. Industry and the environment movement divided like two opposing armies into separate camps. How far this is an accurate picture does not really matter. Certainly it is grossly simplified. The point is that industry is still suspicious of the environmentalist who talks about boosting output and profits through conservation.

What is crucial is that until all sectors become convinced, by whatever means, of the need to introduce conservative-oriented technologies we will make little headway in reaching our aim of managing resources wisely. I believe that we in the environment movement missed a great opportunity in the late 1960s and early 1970s to win industry to our side. We went for the stick rather than the carrot; instead of concentrating on the incentives we were too vocal for the punitive approach.

But in recent years we have been making headway. UNEP, OECD and other organizations have been subjecting the environment to cost – benefit analysis. Though the state of the art has considerable room for improvement, we have been able to demonstrate that nations with environmental programmes in place can make net savings in GNP, that jobs are created, that control measures have been an insignificant cause of inflation, and so on.

Thus, if pressed, we can now couch the benefits of environmental improvement in hard cash terms. For example, in 1979 the US Council on Environmental Quality showed that, against an estimated annual cost of US$19.3 billion to comply with air pollution regulations and air quality programmes, the benefits of measured improvement in air quality since 1970 could be valued at US$21.4 billion per year. Of this total, US$17 billion represented social benefits, including reduction in disease. US$2 billion reductions could be made in cleaning costs; US$700 million could be made by increases in agricultural output; US$900 million could be saved by prevention of corrosion, and US$800 million could be made in increased property values. By 1985, in the USA benefits from water pollution control could be worth US$12.3 billion annually, compared to an estimated cost of US$20.2 billion.

Most encouraging of all has been industry's growing awareness that profit margins can be increased through waste recycling and the voluntary introduction of resource-saving technologies. For example, new processes for sulphuric acid manufacture have reduced emissions from 17.5 to 3.5 kg/ton of sulphuric acid. In the pulp and paper industry recently constructed mills have reduced liquid effluent discharge from 180 to 70m^3/ton of pulp. Other industries have reduced their consumption of water and improved water recycling. In some cases profits have been increased by as much as 40% as a result.

Industry has begun to look upon pollution as a resource in the wrong place at the wrong time. In some countries, industry has shown commendable initiative in switching and then utilizing wastes. For example, the Japanese cement industry is currently using 35 million used car tyres per year in the cement-making process. By burning those tyres, 2 million kilolitres of imported crude oil are being saved every year. Many big companies have now added divisions to provide environmental goods and services. They include Boeing, Exxon, Dow Chemicals, Shell, BP, Krupp, Philips and ICI.

The cost–benefit approach has demonstrated that the strategies which bring the greatest economic return are the preventive ones. The 3M Corporation was one of the pioneers. Its 'Pollution Prevention Pays' (PPP) programme, which is based on conservation-oriented technology, has been a great success story. Between 1975 and 1981, the PPP programme produced total savings of US$97 million; to make an equivalent amount in earnings the company would require US$300 million of additional sales per year.

Thus the new direction for the next decade is the development of recycling and preventive technologies. And by no means is this a message for the industrialized nations alone, for in countries as different as Belize, PR China, South Korea and Brazil, UNEP has logged instances where industrial performance has been dramatically enhanced by the introduction of conservation-oriented technologies. The rapidly industrializing countries have an important chance not to repeat the North's mistakes. Here they should beware the salesmen who want to sell them outdated plant which, though initially cheap, will prove much more costly in the long run.

Governments, of course, do not exist to make decisions on the basis of economics alone. Governments, for example, can make no compromises on the question of human health. An impressive roll call of evidence can now be produced to show that lead from car fumes can cause brain damage in children. The day when the lead content in petrol is universally banned cannot be far away. It is for industry to read the signs, not only on lead but on sulphur, chlorofluorocarbons and other pollutants, and introduce new technologies.

Some industries have benefited from such anticipatory strategies. The outstanding example, of course, was the Japanese car industry which foresaw the advent of stricter emission controls and, in so doing, produced a new generation of low-fuel consumption and low-polluting cars.

But, though increasingly numerous, the instances of introduction of non-waste and low-waste technologies are still the exceptions which have not yet proved the rule. It is for governments through incentives, and industry at its own initiative to step up the production of such technologies over the next decade.

Energy

Only a latter-day Rip Van Winkle, who had slept through the 1970s, would now question the need for radical new directions in energy development. But here too I have a suspicion that in developing new energies we are being mesmerized by the technology itself. Relative to the billions that are poured into nuclear power, into fusion, into maintaining uneconomic coal mines, we ignore fuel conservation. Energy conservation will buy nations time while they develop new technologies and make decisions about the most suitable energy options.

Despite the energy crisis, between 1970 and 1979 world energy consumption rose by a third. Today global energy consumption stands at around 9 billion kW. Industrialized countries dominate the commercial energy market, accounting for about 85% of world consumption. Per capita consumption is about 15 times that in the developing countries. West Germans, by turning down their central heating thermostats to save on fuel bills, managed to save in a year more commercial energy than Kenya used over the same period.

Some progress has been made in energy conservation. Several countries reduced their ratios of energy use to GNP by over 10% without impeding econmic growth. In FR Germany between 1973 and 1980 the GNP grew by 20% but energy consumption rose by only 2.7%. Better insulation is the most accessible of all conservation technologies. Insulation systems now available in the West, for example, can cut household heating costs dramatically.

Side by side, nations must develop new technologies to exploit the new sources of energy, especially renewable ones. A recent study has shown that the potential of new and renewable energy available in Western Europe is enough to meet current demand. Renewable energy technologies, such as solar cells, biogas plants and wind power, have advanced and are advancing fast. But in a sense this is the easy part. The real challenge is to make them available through an effective marketing system. Further, any new technology development strategy should have, inbuilt, a programme for assessing social acceptability.

We have to keep in mind that in the Third World, the need is for cheap, durable and easily repairable hardwares. The choice is not necessarily between candle darkness and an electric light as a cheap calcium carbide lamp can be an intermediary, or between water carried in a bucket or pumped in metal pipes as bamboo pipes can be a temporary alternative, or between literally nothing and a flush toilet as an earth closet is an accessible option. Once again, this should not mean that we lose sight of the ultimate goal of developing countries, ie to stop being treated as second class citizens.

Agriculture

In 1979, world food production stood at 147% of the 1961/65 output, a 10% overall growth in production per head. But can that impressive performance be sustained? In the 1980s there are some disturbing trends. In Latin America and Asia food production is now struggling to keep pace with population growth; in Africa it has begun to fall behind. At a conservative estimate, there are 450 million people chronically undernourished.

The daunting task facing agricultural scientists and technologies is to help devise ways and means not merely of adequately feeding those alive today but of feeding an estimated 2 billion extra passengers expected to board our global aeroplane by the turn of the century. In order to accomplish this, according to FAO, agricultural production will need to be increased by 60%.

The area of land surface suitable for farming is just 11% of the total, some 14 million square kilometres. According to FAO, although an additional 45 million hectares will come into production by 2000, we must contemplate a net loss of 55 million hectares, chiefly through soil erosion and desertification. At present, 20 million hectares are being reduced to zero productivity each year. It is estimated we are losing roughly 25 000 million tonnes of precious top soil each year. The soil loss problem is man-made. It is not climate or other natural factors which are the problem, it is bad management or no management at all.

While we must begin new agricultural development schemes, I believe that not nearly enough money and effort goes into maintaining the systems already in use. Thus, for example, according to the latest Club du Sahel report, as fast as new land is going under irrigation, existing irrigated land is going out of production. This pattern is being repeated on a world-wide scale. Part of the reason is that all too often new schemes have not built in the need for long-term capital infusions.

Lesotho is a country which has experienced devastating soil loss (100 000 hectares, more than 6% of its prime arable land over the last four years). A recent UNEP report found that this was not due to any lack of awareness in government circles of the need to conserve soil or indeed of the know-how to carry out programmes. Soil conservation projects have been undertaken since the earliest colonial days. But the report found that many schemes have broken down through neglect. The money has simply not been there to keep them going. This, of course, is not the only cause of the country's problems; there are other deeper reasons which simultaneously have to be tackled. Some, like land reform, lie beyond the scope of this address.

One important priority is to devise technologies which make the best sustainable use of land already under cultivation. Alien cropping patterns, utilizing alien technologies, have been superimposed on villagers. Take the tractor and plough. Most of the tractors imported are built in Western nations where their design has been dictated by the requirements of temperate zone cultivation. Frequently the ploughs are too deep for the generally more fragile soils of the tropics. In India and Swaziland, two new tractor designs have been manufactured to suit the special needs of the tropical farmer – they are one quarter the price, easy to operate and maintain, low on fuel consumption and made as light as possible so as not to compact the soil. Surely such technologies which are designed with the user in the Third World in mind are the way ahead. Not only in agriculture but in most other sectors, there has been too much uncritical copying of Western models.

In devising technologies to handle the harvest, the first priority should be conservation. Post-harvest losses are a form of waste which could so easily be reduced. There are estimates which show that in South-east Asia 37% of the rice harvest is lost in the journey from the paddy field to the consumer, and that in the USA nearly one third of the vegetable crop is lost. FAO calculates that 10% of the world's cereal harvest is wasted each year. Even if this was reduced by just 2%, 22 million tonnes would be saved – enough to feed 60 million people. Once again, preventing these losses is a tendency easily within our grasp. It can range from something as simple as constructing a crop storage crib to keep crops dry and removed from rodents, to a major national programme to establish rural food processing plants to prevent losses during transport.

Not only in agriculture but also in many other sectors a strong case can be made for making much more use in modern development of traditional systems finely honed over the centuries to the environment. These systems are breaking down rapidly and the old ways forgotten. This is another form of waste. Some West Asian countries are now looking at ways of reviving the Hema, an ancient Bedouin pastoral system which centred on a complex community-operated system of pasture rotation. In Syria 1.5 million hectares are now managed by Hema cooperatives.

Like the Hema, the traditional system of irrigation – called the falaj – in Oman depends on community cooperation. But the intricate falaj system is also labour-intensive, requiring constant maintenance. In Oman there has been a marked drift from the countryside to the towns, with the result that the old system is giving way to mechanical water pumping and overhead water spraying. But modern methods are rapidly exhausting the shallow water-tables which the falaj system utilizes sustainably. Agricultural land in some areas is becoming salinized and hard pans are forming. This is a prime example of the inappropriateness of translated Western technology. Would not the best course be to use modern technology not to replace but to improve such tried and trusted methods?

In many of the semi-arid regions of the Third World, the advent of modern development has resulted in a combining of the worst of the old practices with the environmentally destructive ways of the new. In northern Kenya a UNESCO team of researchers has identified this unhappy combination as the reason for the desertification which now threatens to bring famine to the pastoral nomads of the region. They have produced a management plan which, if implemented, would turn this process on its head. The plan would retain nomadic practices, but combine them with modern marketing, veterinary services, destocking programmes and a system of reserves to be used only in times of drought-induced emergencies.

The haemorrhage of soil loss has to be a great extent been masked by the intensive use of fertilizers. During the 1970s the application of fertilizer rose from 69 million tons to 107 million. But fertilizers are made with fossil fuels and the energy costs are soaring. Already such fertilizers are beyond the reach of even comparatively well off farmers.

The Green Revolution and similar programmes based on a technological package of highly specialized crops requiring heavy infusions of fertilizers and pesticides were a product of the more optimistic 1960s. Increasing pest resistance and the oil price hikes have made them a dubious legacy. In the 1980s we need a new technical package based on nitrogen fixing crops, soil conservation, agroforestry, renewable energies, integrated pest management and practical, low-cost hardwares. A heavy burden will fall on the crop geneticists who must produce new, more productive strains, resistant to pests and requiring little or no fossil fuel-based chemical fertilizers.

Human health

Here the three major challenges for science and technology are the provision of safe water and decent sanitation, the production of safe chemicals and the safe handling of hazardous waste. In their book *Only one Earth*, Barbara Ward and Rene Dubos singled out lack of access to safe water and sanitation as the first threat to the peoples of the developing world. Surely one of the greatest indictments we can make of the values we attach to the application of science and technology has been the failure to provide safe water? Once again, we find that a revolution is required not in technology but in our attitudes.

In 1983, one in four people living in the cities of the Third World have no access to clean water. In the rural areas the situation is even worse, where two people out of three must drink and wash with dirty water. In the cities 50% are without decent sanitation; in the countryside 87%. Contaminated water is the carrier of diseases like typhoid, cholera and polio, which kill millions each year and which debilitate millions more. Reaching the UN target of 'clean water for all' by 1990 will require the commitment of US$300 billion – a mere fraction of the sum the world will spend on arms or even tobacco. Putting projects in place to meet that target lies comfortably within the reach of our technological capacities. Often it involves something as simple as the construction of a standpipe in a slum, or the installation of a concrete lip on a well head.

It also lies within our compass to defeat other environmentally linked diseases. Poorly planned water development projects have helped maintain bilharzia which strikes 200 million people in the Third World each year. In the 1960s we talked confidently of eradicating malaria. Across the tropics malaria is as common now as it was 20 years ago. In some countries, most notably India, malaria has been making a strong comeback. River blindness and sleeping sickness are other environmentally-linked diseases which also lie within our ability to eradicate, or at least minimize.

People's health is also being put at serious risk by the increasing volume and numbers of potentially dangerous chemicals. WHO estimates that half a million people are poisoned by pesticides each year. In the use of pesticides and other harmful chemicals, one very disturbing trend has emerged over the previous decade: with the industrialized countries evolving stricter standards of safety, we now have one standard for the rich and one for the poor. But a chemical like DDT is harmful on whichever side of the global economic divide it is used. Nor do these dual standards ensure the immunity of the industrialized nations. Food imports from the developing countries have been found to contain dangerous quantities of harmful pesticide residues. For example, the US government found recently that 50% of all imported coffee beans contained pesticides banned in the USA.

Science and technology have a crucial role to play in producing one standard for our one world. The need is to strike a balance between safety and the requirements of development. If safety tests become too stringent, too protracted and consequently too expensive, many potentially very beneficial chemicals could be kept off the market. UNEP, with its International Register of Potentially Toxic Chemicals, is endeavouring to create such a balance.

People's health is at risk not only in the production but also in the handling, transport and disposal of hazardous wastes. Again, we need uniform and globally applied standards. Public concern has been expressed forcefully over the random and poorly controlled dumping of poisonous wastes on land and at sea. A recent US Environmental Protection Agency survey logged a total of 181 000 lagoons and assorted water impoundments for industrial waste. Seepage from these dumps presents a serious risk to underground water supplies.

The challenge is either to produce technologies which eliminate waste at the manufacturing stage or to recycle it. Where this is not possible, a means must be found of rendering the end product harmless. Top of the list must come attention to devising means for the safe disposal of radioactive wastes.

Towards a new ethic

We have, then, the power to design a better future; to use science and technology not only to safeguard but enhance the quality of our shared environment. We know too what the major challenges are. But in order to take the new directions I have described, we need first to transform the values that we attach to science and technology.

I have heard calls for the need for a new morality, a new ecological ethic. The transformation to a fairer, less wasteful, more conserving order does, indeed, demand a new ethic. But it will not happen through rhetorical appeals, it must be accomplished almost without realizing it.

I have seen recently an appeal by the Human Rights group Amnesty International for support. The appeal was signed by leading celebrities who put their names to an advertisement asking for Amnesty to be eliminated. Amnesty, said the advertisement, could be made redundant overnight if every nation applied the principles of the Universal Declaration of Human Rights. How much the same applies to UNEP and the environment movement as a whole, were the Stockholm and now the Nairobi principles to be applied.

I would like to see UNEP become redundant. I would like to see a day when automatically, the UN, the development assistance agencies, governments, businesses, farmers and others will take the environment into account at every level of decision making. Nor is this an impossible dream. In essence we are asking merely that nations – governments and people – look upon the living resources of the biosphere as something of measurably more value than finite resources. These resources are our environmental capital, and we still have enough capital left to live off the interest. If we cannot apply a value system which employs science and technology in the service of using these resources prudently and fairly then we will, indeed, be on course for a global catastrophe.

The environmental dimension of economic development has won wide acceptance at the conceptual level. Everyone, from the farmer and the factory worker to the government minister, must now see the concept work in practice, to see that environmental management brings both short- and long-term benefits.

But the environment can never be reduced solely to profit. It can never be viewed exclusively in pragmetic terms. The physical health of people depends on healthy surroundings; on this, as I have said, there can be no compromise. People's cultural and other aesthetic needs, too, offer no scope for trade-offs, and spiritual needs similarly.

These practical and unquantifiable needs are not in any way mutually exclusive: they reinforce each other. Take the question of wildlife and genetic resources. In my own country, Egypt, a herbaceous moisture-loving sunflower plant which was once quite common on the banks of the Nile appears to have become extinct. The stems have been used by weavers since ancient times for making baskets, like the one which carried the baby Moses. So it had a practical use, but the sunflower also had an unquantifiable cultural value. Who does not feel a loss at the disappearance of any such species, and one that can never be fully explained? Such feelings are the heart of the environment movement. I hope the day will never arrive when our brains are allowed to rule the heart completely.

Rene Dubos said we have 'humanized' the earth. This is true. But we have yet to devise a technology that allows us to live apart from the biosphere. Our challenge is to apply systems that work with the biosphere rather than against it. With their confidence in human innovation, energy and sheer common sense, Rene Dubos and Barbara Ward believed to the last day of their lives that we could apply such systems. They did not underestimate the difficulties, but neither would they allow themselves to be daunted by them. The human spirit can prevail, we can overcome the problems we face today. We have, as Barbara Ward said, 'a duty to hope' that humanity, which has created all the problems I have outlined, will mend its ways.

10

The peril of the two cities

Statement to the Sixth Session of the Commission on Human Settlements

Helsinki, Finland, April 1983

Through the joint UNEP and Habitat bureaux meetings and our collaborative activities, the tradition of close cooperation between UNEP and Habitat continues. I am confident that, through the medium of thematic joint programming, we will see over the next few years a radical improvement in cooperation between UN agencies collaborating in SWMTEP, the System Wide Medium Term Environment Programme. I should add, however, that all agencies are in agreement that the funds earmarked for 1984/85 to implement the part of the medium term programme concerned with human settlements planning are not adequate to meet the needs outlined by governments.

Finland in general and Helsinki in particular constitute rich proof that harmony can be achieved between buildings and nature and between design and social and cultural needs. I hope that the commitment to sound human settlements planning shown by the Government of Finland will serve as an inspiration to those from other countries as they come to grips with one of the most daunting problems facing mankind today – the provision of a decent human environment for the under-privileged two thirds of the world.

In UNEP we see two broad issues concerning human settlements and the environment. The first is the external impact – the way in which settlements create environmental problems by misusing resources such as land, fresh-water and firewood. The second issue is the quality of the internal environment cities provide for their inhabitants.

External impact of human settlements on the environment

The quality of both the external and internal environments of cities are deteriorating alarmingly. As always it is the poor and the underprivileged in the rural villages, townships and big cities who suffer most. Through forces beyond their control, their conditions of life are being dehumanized. For this reason UNEP welcomes the choice of Land for Housing the Poor as one of the two major themes for this Sixth Session of the Commission on Human Settlements.

Some of the large cities of the developing world have already expanded beyond the capacity of the external environment to provide their needs. Supplying cities like Calcutta, Mexico City, Sao Paulo and Cairo with fresh water already presents immense problems; around Sahelian townships the demand for charcoal has deforested huge areas in ever-widening circles; and uncontrolled urban growth is eating into prime agricultural land.

If the major urban areas are already seriously out of balance with their external environment, by the turn of the century – now 17 short years away – they will be grotesquely so; that is if governments and the international community carry on as they are.

Quality of environment within cities

Even external problems are relatively minor, however, when compared to the problems of environmental deterioration within town limits. Here the environment is being ruined, principally through poverty but also through the inability, and sometimes the refusal, of urban and national authorities to tackle its underlying causes.

A new idiom has entered the language – hyperurbanization. The trend has been for a concentration of urban populations in a few large cities which act like a magnet for villagers from the neglected rural areas. These 'megacities' receive the lion's share of public and private investment. This trend is unlikely to change in the foreseeable future. But even if immigration to urban areas were somehow to be reduced to a trickle, we would still have to come to terms with planning for a time a generation or so from now when, if the present trend continues, urban agglomerations of 10, 20 and even 30 millions will become common.

Many of the problems centre on land, and specifically on how we use it – land for building, land for open spaces, land for growing food and fuel crops and so on. One of the most obvious failures in land-use planning has been the loss of fertile land to low-density urban sprawl. World-wide, an estimated area in excess of 70 000 square kilometres was lost to food production through urbanization during the past decade. This particular misuse of land can be halted. Countries should do more to build up indigenous town planning capacities and they should see to it that the property speculators, the only winners in this process, are controlled. However, this cannot happen without a strong political will. When we look across the international political landscape we find that will to be sadly inadequate, or worse, missing altogether. Nowhere is this more so than when it comes to providing a decent internal environment for the underprivileged, one that will provide for basic human needs and the requirements of human dignity.

In a way it is misleading to talk of a city as though it is one entity. In most major Third World urban settlements there are two 'cities': one for the elite where Western standards prevail and one for the poor – the usually self-built cities or, in less polite terms, the slums and shanty towns. Spatially, economically, socially and politically, the chasm between the two cities has never deen deeper or wider.

The slums and shanty towns are concentrations of misery and growing despair which no statistic can adequately convey. These areas are increasingly denied access to clean water, safe sewage and garbage disposal, medical care, adequate housing and energy. No wonder one child in three living in the deprived city does not survive beyond five, no wonder the crime rate is spiralling, and no wonder the poor have begun to stop even hoping for a better future. Where this has happened, apathy and fatalism – the twin threats to action – have set in.

Improving the internal environment of cities

What actions, then, are needed to stop the wholesale deterioration in the internal environment? Of course, no easy solutions exist. But whatever the solutions, they must be based on bridging the divide, on bringing the two cities together. There are powerful incentives for governments to begin the work of bridge-building, for the divide is an artificial one. Environmentally, a human settlement cannot be separated: contaminated air and water and noise pollution are shared; diseases and fire spread; a breakdown in refuse disposal or sewage systems affects everyone; and destruction of natural beauty offends all. When we add considerations of social justice, humanity and political stability, the case for the 'haves' to begin in earnest the work of radically improving the environment of the 'have-nots' is overwhelming.

The first and most obvious way to improve the environment of the poor is to provide a secure and stable economic base. But governments also need to implement programmes to promote efforts by the shanty-town dwellers themselves to improve their own habitat – better sanitation, installation of standpipes, better garbage disposal and recovery, voluntary family planning, hygiene, education and so forth.

Governments, in cooperation with city authorities, could be doing much more within the framework of sound planning to improve the provision of land to the poor by, for example, granting security of tenure on illegally settled plots, or making available cheap loans and serviced housing plots. However, we should not forget that ranged against such reforms will be the property speculators and other entrenched financial interests. Political commitment of a high order will be required to implement national programmes aimed at improving the lot of the poor. Better coordination is required between relevant ministries and more support needs to be directed to city administrations to pay for the mounting of programmes to enhance the urban environment.

Improving environmental conditions in rural areas

The case for reform applies with equal force to the even more neglected rural areas. If there has been a steadily increasing exodus over the previous 20 years from the million or so villages of the developing world, it is largely because conditions in the remote and politically disadvantaged villages are frequently even worse than in the cities. In 1980 it was estimated that 71% of villages lacked safe water supplies and 87% were without sanitary facilities. These grim statistics are in part a reflection of the feeble political clout of the rural areas. Simultaneously, we must step-up the work of improving the environment and economic opportunities in the countryside. Until this is done, governments in the Third World will not be able to stem the flow of migrants which, together with natural increase (running on average at 2.5% each year), are creating the new megacities.

Environmental refugees

There is one further category of underprivileged – the refugee. World-wide there are 10 million refugees who need to be clothed, housed and fed. Providing the basic needs for the legions of displaced people has already placed the economies and ecosystems of the countries of asylum under intolerable strain. Somalia is one such country. UNEP, in cooperation with Habitat, the High Commission for Refugees and the Government of Somalia is preparing to start a pilot demonstration project. Our expectation is that governments and the international community, including relief non-governmental organizations, will take inspiration from this project once it is implemented.

UNEP has begun to extend the concept of a refugee beyond that of a casualty of war and civil strife. There is evidence from East Africa, the Sahel and the Caribbean islands that soil erosion, deforestation and watershed destruction have reached a point where people can no longer scrape even a bare living from once productive farmland and forests. Their only option is flight. We have begun to call these people environmental refugees. But exodus is no long-term answer as it merely delays or shifts the problem. The only long-term answer is to promote rural development which helps villagers to increase productivity by conserving their renewable resources.

The International Year of Shelter for the Homeless

Finally, I move to the second major theme of this Session. The year 1987 will be the International Year of Shelter for the Homeless (IYSH), the initial preparations for and the objectives of which this meeting will consider. On behalf of UNEP, I applaud the Prime Minister of Sri Lanka and the Sri Lankan Government for their commendable initiative and pledge UNEP's wholehearted support for the campaign. The IYSH provides a great opportunity to focus the world's attention on the problems and solutions I have described in this address.

UNEP welcomes the decision of the General Assembly to designate Habitat as the centre to coordinate the IYSH and will extend the maximum possible support to its neighbours in Nairobi as they prepare for 1987.

11

Putting the principles to work

Statement to the Fourth Meeting of the Committee of International
Development Institutions for the Environment

New York, USA, May 1983

This meeting is held against a background of persisting world recession. UNEP fully understands the massive impact deepening debt and a standstill in the funding levels received by members of the Committee of International Development Institutions for the Environment (CIDIE) have had on the Committee's activities. But this does not preclude the need to ask some tough questions as to how far CIDIE members have reflected the principles of the Declaration of Environmental Policies and Principles for Economic Development in terms both of their policy objectives and in practical implementation. Some CIDIE members have not yet even established a formal in-house environmental machinery. UNEP's view is that performance, on all fronts, leaves much to be desired.

There have been a number of very positive developments, but these have merely served to show us how much more might have been achieved and, more important for the purposes of this meeting, how much more could be done. On the positive side, we can cite the Carajas Iron Ore project in Brazil, which includes a comprehensive environmental management programme; the Organization of American States' pioneering work in watershed management; the success of the Gujarat social forestry programme; the Jengka Triangle development scheme in Malaysia which carried out an intensive land-use survey prior to the settlement of 9000 farming families; and the welcome change in the World Bank policy to take account of the special needs of tribal peoples in new development schemes. The list is long, but not nearly long enough.

Against this trend we continue to learn of the larger infrastructure projects receiving little or no attention regarding their environmental impact. We have evidence too that little more than cosmetic adjustments have been made to existing development assistance programmes. Take the Sahelian countries: according to a Club du Sahel report, as fast as new land is going under irrigation, an equivalent amount is lost through salinization; and in the same region we learn that, although tree planting is running at 2% of demand for firewood, only just over 1% of all aid goes to agroforestry. We learn too that funds are still available for unsustainable tropical forest development, despite the lessons we have learnt about the ecosystem's fragility. If forest in the Amazon Basin is being converted to useless scrub growth at the rate of 100 000 hectares per year, the multilaterals must shoulder part of the blame.

We have then a good idea of what sustainable development *is not*, but we are far less clear about what it *is*. We need more thoughtful policy research and more forceful and systematic procedures if we are to succeed, first, in deciding what our environmentally-sound objectives should be and, second, how we are going to apply them.

Dwarfing the funds committed by the development assistance agencies are the private investments in megaprojects. It falls to the multilateral institutions to lead by example; to show that productivity can be increased and sustained by sound environmental management. The merchant banks and industry in general have made too few meaningful commitments to the environment. In addition, we have no capacity for assessing and monitoring their projects. UNEP is negotiating with industry to stage a conference in 1984 with the aim of securing a formal commitment to the environment. At the very least it will provide a standard, a yardstick, by which they can operate.

Priorities for CIDIE members

One priority for CIDIE members should be to finance the preparation of more workable and more cost-effective environmental impact assessments (EIAs). You would agree that, as management tools, EIAs tend to be too long-winded, too time consuming in preparation, too costly to produce and usually couched in impenetrable jargon. Even where the goodwill exists, EIAs have been found wanting. I have in mind the experiences of my predecessor, Maurice Strong, who, while heading up a Canadian oil company, found he had few procedures that were of any use to his engineers, planners and designers. And that was in a developed nation. The situation is gloomier still in developing countries where, frequently, little or no indigenous capacity exists. Typical in many ways was a well designed World Bank ore mining project in Peru. On paper it met all the requirements, but proper observance of the environmental component depended on costly and time-consuming visits by expatriate experts. In Papua New Guinea a multimillion dollar gold and copper mines project will soon be underway. The company concerned has promised EIA assessments but, even if they are forthcoming, the country lacks the capacity to evaluate them.

In developed countries such as the USA, EIA statements account for approximately 1% of the total cost of projects, as compared to the 10% usually allocated to planning. But even 1% may perhaps be too high for many Third World countries. We should examine seriously ways and means to reduce those costs to fractions of 1%.

Though we can claim to have made fair progress in educating supply, we have made little impact on demand. Desperate for investment, developing countries will often say yes to private investment schemes and pay little heed to the long-term impact. CIDIE members, through example and through assistance in many forms, can help these countries to be more discriminating. One important step should be for the CIDIE members to establish environment budget lines. Such a development would show recipients that much more than a token commitment to sustainable development is being made.

The development assistance agencies, both bilateral and multilateral, and the environment movement could be doing much more to coordinate their work. Sometimes from several different sources, the developing countries are receiving help on national environmental data supply, on environmental profiling and so on. For example, Sri Lanka is currently being supplied with national conservation and environmental protection plans by five different organizations. I am not calling for a single monolithic approach, but we should at least be aware of one another's plans so that precious manpower and resources are not wasted on duplicate projects.

UNEP needs the help and cooperation of CIDIE members in its programme of environmental cost–benefit analysis (CBA). We have found that we can invoke economics to show that environmentally-sound development can increase profits and productivity. But we would be the first to concede that CBA is at an imperfect stage of development. We need, for example, to produce airtight CBA analyses which take account of traditionally non-economic factors such as health and improved quality of life brought about by an enhanced environment.

To recap, my concrete recommendations are that, as a matter of urgency, CIDIE should consider:

- stepping up efforts to build up local capacities through education, training and research, which will have the effect of helping developing nations keep economic development on sustainable lines;
- helping to produce more cost-effective EIAs which will be presented in simplified and, where feasible, standardized formats;
- helping to improve radically the state of the art of environmental cost–benefit analysis so that it serves more closely the needs of economists;
- improving coordination with respect to collection and presentation of environmental data of recipient countries to avoid duplication, which is likely to occur during the preparation of environmental profiles, national conservation strategies and environmental chapters in the policies of development aid agencies;
- improving coordination with those interested in the environment in the development community.

More attention should also be given to financing projects aimed specifically at helping Third World countries deal with their most serious environmental problems. In this regard, UNEP's Governing Council has given the go-ahead to a clearing-house type of facility which will cover the identification of chronic problems in selected countries and devise programmes of activities to cope with them. UNEP is open to advice and guidance on how to proceed. To this end I have invited the presidents of CIDIE institutions and heads of multilateral financing bodies and bilateral aid agencies to a meeting in Geneva in early July to give them a more detailed briefing on the clearing-house programme and projects.

Performance of bilateral aid agencies

Despite many positive developments, UNEP continues, too, to be disatisfied with the performance of most bilaterals. We are aware, of course, that each operates from a different set of policy directives reflecting defence and foreign policy objectives, trade flows, multinational networks, transport and communication arrangements and so forth. Including these requirements in a Declaration of Principles would seem a near impossible task. We feel the best way to proceed is through the development of a set of common aid criteria for donor countries. The mechanism of the Development Assistance Committee of OECD could be used for this purpose.

Unfortunately, there is very little hope that anything meaningful will be done unless, and until, the tools and methodologies for integrating environmental concerns into projects and programme development are not only adapted by each institution for its own purposes but willingly accepted as essential analytical apparatus. The clearing-house type of facility will produce a set of model projects where the environment–development interactions have been taken fully into account. The CIDIE members and bilateral aid agencies should give close attention and, if possible, support to these projects and programmes until they become self-supporting.

Involving local people in project design

I would like to make two further and connected points. The first is that far too little account is taken in project design of the need both to involve local people and to tap their already existing respect for, and knowledge of, the environment. Development is not only science and economics, it is also aesthetics, sociology and anthropology. How often before we embark on multimillion dollar projects do we find out how the people most concerned feel? What use are modern slaughterhouses if pastoralists in remote regions are, for reasons of status, reluctant to sell their animals? What use are nitrogen-fixing beans if local villagers will not eat them or do not have enough firewood to boil them? And what use are more efficient types of cooking stoves if they do not produce enough smoke to de-infest the thatch roof?

It is on these seemingly minor criteria that the success of a project so often depends. In the Ethiopian Highlands, against an alarming national decline in productivity, in one small 250 square kilometre catchment area productivity is going up – at least enough to keep pace with population increase. One reason is that the project, supported by the World Bank, was careful to find out what the villagers' needs were before embarking on a programme based on fuelwood plots, double-cropping, soil conservation and so on. The other was that the experts had been prized out of their pigeon-holes to evolve an integrated project. No massive infusions of cash have been made – aid in this case has come mostly in the form of advice and small-scale technical assistance.

Blending advanced technological development with traditional knowledge

My second point is the neglect of traditional knowledge of the environment. I have in mind, for example, the Asian agroecosystems, such as the Indonesian 'Sawah' method, which have served the people so well over the ages. How often do we begin with the premise of using such systems as the starting point and then seeing how modern methods can be included to improve productivity? It is surely no coincidence that in Syria and Saudi Arabia they have begun to look at ways of reviewing the Bedouin Hema cooperative system which is more than 15 centuries old; that in Iraq they have reverted to age-old irrigation techniques to salvage the salinized Greater Mussayeb project; that in Oman it is now suggested that the centuries-old 'afflaj' drainage system be restored.

I am not advocating a Luddite solution, that is a return to a preindustrial economy. Rather we should apply alternative and less wasteful patterns of development and lifestyles which revive a respect for living resources. Simultaneously the development aid agencies should do all they can to bring the maximum material benefits of the most advanced technological developments through blending them with traditional wisdom – that is, combine the best of the old with the best of the new – to increase living standards. Surely leadership by example is the only way forward.

I should add too that in UNEP we are not expecting CIDIE members to put the environment before all. However, we are expecting them to do more to improve environmental management; which is another way of saying that we must ensure that new schemes maintain productivity.

Conclusion

I am aware that my brief address may seem to have simplified what are a series of very complex issues. This has been deliberate for surely one of our main tasks is to demystify environmental procedures and, in so doing, make them more palatable and more useful to the engineers, the bankers, the farmers, the industrialists and the planners on whom, in the last analysis, we depend if we are to reach our goal of sustainable and equitable development. The Declaration has been, for the most part, a paper tiger. It is the job of CIDIE members to give it teeth and claws.

12

Towards a more prosperous global tomorrow

Statement to the Global Tomorrow Coalition Conference

Washington, DC, USA, June 1983

The theme of this meeting is rebuilding US leadership. For more than 10 years I have observed the evolution of the USA's role in the world environment movement. Concern has been expressed in the press, in NGOs and other fora, that the USA has appeared to relinquish its pioneering role in addressing environmental problems. That concern has surfaced in the world's press – from London to Bombay, from Caracas to Tokyo – where terms like 'backsliding' and 'environmental isolationism' have appeared in print. The USA's openess and prediliction for intense, public debate is a way of conducting national affairs that other societies do not sometimes readily understand.

They may not readily understand also a conference convened to restore US leadership – they might be forgiven for thinking you believe they are only waiting for the USA to give them a lead. However, we know that the purpose of this gathering is to identify ways and means the USA can, first, secure the many positive advances it has made and, second, build on those developments.

In the international community we are looking to the USA to apply more thoroughly an understanding I know it holds deeply – that as a good neighbour in a global neighbourhood, it sees its own well-being in the light of that of its neighbours. For evidence of such an understanding we need look no further than the agenda for this meeting. The significance of all the environmental problems on the agenda – acid rain, the skewing relationship between population and resources, hazardous exports, the health of the oceans – go well beyond geopolitical divisions.

I believe the expression is that when America sneezes the world catches cold. What should be added is that America can also catch cold from others. It is already happening. Take hazardous exports. Over one half of the pesticides used in the Third World are applied to crops destined to be exported to the USA and its OECD partners. The Federal Drugs Administration and other US agencies have found dangerous pesticide levels in imported foodstuffs such as beans, peppers, beef, tea and coffee. UNEP's 1983 State of the Environment Report estimates that 20% of all new chemicals coming on to the market each year are potentially hazardous to human health. This threat to health may be posed through numerous pathways.

The dwindling of the earth's biological diversity poses a similar global threat. Not only in the USA, but throughout the world, agriculture, pharmaceutical and other industries are being undermined by the accelerating pace of animal and plant extinction.

I am conscious, of course, that this sophisticated audience is fully aware of such shared perils, and also that global documents such as the World Conservation Strategy and the Brandt Report have cogently argued the case that environmental impoverishment is undermining sustained global development, and in so doing posing a major threat to the peace and security of all nations.

But we have to recognize that outside this meeting, in many sections of government, in industry and commerce and other sectors, more often than not these threats are dimly perceived. Even when they are readily perceived, decision makers are frequently at a loss to know how to take preventive steps and apply the remedies.

US leadership and global interdependence

Against this backdrop the world is, indeed, looking for US leadership – a new kind of leadership which is based on a comprehensive understanding of mutual self-interest. We are looking to the USA to provide a new kind of leadership finely honed to the requirements of global interdependence.

The whole world is sharing in economic hardships. In these times, there is still much that US leadership can provide. Leadership does not always mean funds. It comes in many forms. The USA can still provide considerable assistance by supplying scientific information and expertise. It can cooperate in the evolution of international environmental policies and conventions. Unwavering adherence by the USA to its long-standing environmental principles and policies is crucial to the credibility of US leadership.

The USA has a firm foundation from which to start. Its innovative government procedures have been, and continue to be, a source of inspiration: eight major national laws encompass concerns from air and water quality to protection from noise and radiation; at all levels, the judicial system has reaffirmed the integrity of environmental legislation; few nations can rival the support given to environmental education at state and federal levels; the placing of senior officials with an environmental monitoring brief in all federal executive departments has not been emulated. Some countries, including Nigeria and PR China, have sought, and received, help on technologies for clearing the air and guidance in establishing national environmental agencies. Still others have used US regulations as a blueprint for their own, for example, the USA's 208 and 404 water regulations are cited in Japan. The US National Environmental Policy Act has provided a guide for developed and developing countries alike.

We are looking to the USA to build on its initiatives and achievements to date in two particular areas. The first concerns the pioneering role of the USA in environmental cost–benefit analysis and application. I am convinced that in this time of world economic strain we will fail to get the environment included in the mainstream of economic decision making unless we go armed with hard and fast evidence that conservation and other aspects of environmental protection can boost productivity and enhance profits. Industry and government have begun to provide that evidence: 3M and Dow Corning are only two US companies among a growing number to have discovered that recycling through a 'systems approach' of integrating one kind of processing plant with another can result in considerable savings. One reliable survey has found that environmentally-induced economic activity can stimulate the economy by as much as 2% of the GNP – a case in point that what is good for an individual company is good for the economy as a whole.

But the state of the art needs to be advanced more rapidly. There is massive scope for US leadership here. There is a great need to formulate ways in which we can utilize cost–benefit analysis in addressing environmental problems. The second area where we are looking to the USA for inspiration concerns overseas development assistance. The USA, which disburses about US$8 billion each year in official development assistance, has already played a leading role in promoting sustainable development. USAID was a trail-blazer in establishing environmental impact procedures, in preparing environmental profiles of nearly 50 nations and in investing well over US$100 million annually in environmental projects.

The 1980s are opening up a host of new opportunities to build on that record, and not only in the bilateral sphere. The USA could be doing much more to use its muscle with the multilateral agencies to reinforce their commitment to the environment. For example, according to the latest Club du Sahel report, the coalition of donors established after the Sahelian drought, a tiny 1.3% of all aid goes into agro-forestry, while over 30% is still in the form of food hand-outs, which although absolutely necessary at times of food shortages sometimes have the negative effect of creating dependence. With the NGOs in the forefront, much more could be done first to expose and then to promote ways to redress such imbalances in development assistance.

The USA could be doing more in helping to make environmental impact assessments better tools for developers. An urgent priority is building up indigenous capacities through education and training; and more assistance should be directed to helping developing nations deal with their most chronic environmental problems. The list is long.

It is against the background of solid achievement and a better understanding of global interdependence that I should register my surprise that the USA has not endorsed the Law of the Sea Treaty, and that it should have felt it necessary to cast the only negative vote among 147 nations on the UN initiative to prohibit export of banned hazardous products. We are concerned too that the USA seems to be alone among Western nations in moving away from imposing tougher controls on sulphur dioxide emissions. But against this trend we noted the valuable contribution the USA made at our most recent Governing Council in Nairobi and at Cartagena where the USA became one of the signatories to the treaties under the Caribbean Action Plan.

I understand that we must recognize that in the evolution of any public policy there is time needed for stock-taking: to pause, to refine, to reassess the courses chosen and the goals defined. There is always a need to consolidate past gains if further progress is to be durable. Emotion and commitment cannot substitute for facts, for hard evidence. Grand solutions applied at one level may create unforeseen problems at another. It is not dishonourable to take stock if that is the true intent. UNEP honours that process. But you cannot read the world's newpapers without wondering what road the USA might take.

Some fellow environmental travellers are surprised that, after a decade of confrontations within the USA between various sectors of society, the debate has begun anew.

The debate at Stockholm in 1972 between the developed and less developed nations as to whether or not the environmental agenda was even appropriate has long-since been silenced. The world community has achieved a significant consensus that safeguarding the environment is a fundamental goal of all nations. We 'environmentalists' no longer have to provide the burden of proof. Sometimes harsh experience in the form of soil erosion, silted dams, deforestation, corrosion and so forth has provided us with the proof we would prefer not to have.

Learning from the experience of others

National debate in the USA must not stand in the way of an international commitment. For the USA has both much to contribute and much to learn from the experience of others. A humbleness and a willingness to be inspired by the example of others is central to real leadership. Take acid rain. The USA should take full note of the steps its Western European partners have taken or are preparing to take to combat the acid fall-out which, according to our State of the Environment Report, is affecting up to 4 million square miles (10 million square kilometres) in Europe and North America. FR Germany, for example, will spend up to US$5 billion over the next 10 years on tackling the problem. New cost-effective control techniques are being pioneered. Sulphur emissions have been dramatically reduced by 'fluidized-bed combustion' and other techniques. We should take advantage of the revolution in information technologies to ensure that news of these developments becomes readily available.

The USA must be instrumental in building an even stronger scientific and legal base, to demonstrate institutional models by which the many and continuing questions can be assessed. Are regulations more or less effective than incentives? Are your regulations designed in the best way? What laws have proven to be the most cost-effective; what technologies, what institutional procedures? Who is best placed to finance environmental measures? With these, as with many issues, the response of the USA may hold important keys to the world's continuing quest for cost-effective environmental solutions.

An area of vital concern is exploring the potential for the environmental organizations and the corporate sector to work more effectively together. Thus, to my model of leadership I should add reconciliation. Not without good reason, business in the USA and abroad is sceptical about our motivations. Business still tends to underestimate the benefits of protection and, perhaps, in our zeal, we tend to overestimate benefits. Our endeavour over the next few years should be to bridge the gap – to go more for the carrot and less for the stick.

Most of you will be familiar with the report exposing the 'hamburger' connection in the destruction of Central and South America's tropical forests. Reportedly, less than 2% of US beef imports come from that area. Often wholly ignorant of the fragility of the moist forest ecosystem, private US investment has gone into setting up short-lived cattle ranches. What is happening is serving no one's interests, least of all the private investors. This is but a microcosm of a global problem – we must martial the facts and the data at our disposal to convince firms that investment which takes little or no account of the environment is a futile, loss-making exercise.

Public attitudes to environmental protection

The ultimate responsibility for leadership lies with the individual citizen. The effectiveness of environmental protection depends in the last analysis on the public's attitude. It falls mostly to the NGOs to guide the public's attitude. By organizing and articulating the environmental aspirations of the citizens of this country, US NGOs have provided the backbone of US environmental action.

General warnings and exhortations are not good enough: we need in the environmental community to show what can be done by individual citizens acting alone and together. The public opinion specialists will always tell you that people will respond much more to a crisis when they are given clear indications of what they can actually do about it. In the UK a campaign modelled on US experiences to alert citizens to the dangers of lead in petrol has been remarkably successful. A series of rallies, meetings and a complementary press campaign has wrung from the government a commitment to a timetable to eliminate lead completely.

UNEP needs the full participation, interest and vision of the NGOs at the international level to support, augment, prod and, where necessary, criticize national and international initiatives. NGO support helps create world pressure for action. The ever-increasing knowledge and participation of NGOs in the development process is crucial. Never underestimate the power of example, and do not forget that in societies far less open than the USA, public opinion and pressure is also effective in influencing decisions.

UNEP, serving the whole world, would serve poorly if it did not work in close and creative partnership to promote the agenda for sustainable development. For this we need your creative thinking and supportive leadership.

Our intention is that during this decade, UNEP will play a conciliatory role among interest groups, and among nations and the public and private sectors in dealing with issues. But member governments must be committed to the mandate we were given. Our effectiveness is dependent on the political will of member states, and NGOs play a critical role in creating and sustaining that will.

We have seen increasing responsiveness from the Bureau of International Affairs in the US Department of State. We welcome old friends at the top echelon of EPA. Members of the US Congress continue to play a pivotal international environmental role, not only in support of UNEP, but in their many other endeavours. And USAID continues to serve as a beacon of light.

For its part, UNEP will consider seriously the model of leadership that will emerge from this meeting. Our hope is that it will be based on reconciliation, on pragmatism and on an understanding of mutual self-interest. The time when the USA could consider itself an economic and environmental fortress is over – in short, your important task is to show how the USA, in a world of interdependence, can contribute to our reaching a more just and a more prosperous global tomorrow.

13

Profiting from the environment

Address to the Centre for World Development Education

London, UK, October 1983

The term 'human environment' is an all-encompassing one, signifying the totality of our surroundings and our relationship with those surroundings. In UNEP our main preoccupation focuses on how wisely, and how fairly, mankind uses the earth's resources, both finite and renewable.

In my discussions with representatives from business and industry I consistently put the case that UNEP's preoccupation should be industry's too; for the range of resources needed to be conserved has increased greatly. The time has come when we must look upon clean air, water, topsoils, wild plants and forest cover as resources every bit as precious as, say, coal, oil or mineral ores.

We have traditionally taken these self-regenerating or 'renewable' resources for granted. However, these resources are renewable only if conserved, and are finite if not. Through a combination of avarice, irresponsibility and ignorance we have in effect been mining these sustainable resources; we have been paying the price in the form of resource depletion.

Within the developing countries avarice – by which I mean overconsumption and the pursuit of wasteful lifestyles – is a minor cause of environmental degradation relative to poverty. Most of the poor nations of the Third World are locked into a cycle of resource destruction and deprivation. Handicapped by population increase, unfair terms of trade and initial poverty, the governments of these countries have little scope for preventing their poor, who live on the margins of existence, from destroying the very resources on which their future livelihood depends; a process which was accurately described recently as one of 'self-immolation'.

Role of the environmental factor in world development

But against the rather bleak scenario described above I can report that the environmental factor in world development has begun to play an increasingly important role; we have made rapid headway among the decision makers who work in government, the UN agencies and the development assistance agencies. Few, if any, in those cabals would put a contrary case that development which takes no account of the environment is worth pursuing. At the launch of the World Conservation Strategy, the then UK Environment Secretary, Michael Heseltine observed that 'in any individual decision the starting point will be conserve what matters – those who have a contrary objective must bear the onus of proof'.

Perhaps the most obvious sign of this transformation is a change in the environmental vocabulary: the emphasis is no longer on 'small being beautiful' or on 'limits to growth', instead we talk exclusively in terms such as sustainable development and rational resource management. These are not new catch-words, but, significantly, none have come along to replace them. The overwhelming concern of the responsible environment movement nowadays is to see these concepts put to work. The world now desperately needs skilled manpower and money to move the ideas of sustainable development into meaningful action.

Aid agency assistance

The multilateral development assistance agencies which spend, between them, more than US$15 billion on new projects each year, have radically altered their lending policies in favour of the environmental factor. The turning point was the signing in 1980 of a Declaration which formally committed the World Bank, the UNDP and the other major lending and development financing institutions to environmentally-sound development. The purpose of the Declaration was perhaps best summed up by World Bank President Clausen when he said in a major public statement that 'sustainable development and wise conservation are, in the end, mutually reinforcing and absolutely inseparable goods'. Since then, several country-to-country aid agencies, including those of the USA, FR Germany and Sweden have made a similar commitment.

Private business and industry investment

But the amount disbursed by the multilateral and bilateral organizations pales into insignificance when compared to the investments industry itself and the private financing institutions make each year. I have no mandate to speak on behalf of our partners in the environment and development community, but I believe they would not disagree with our view in UNEP that, in comparison, our impact on private business and industry has been small indeed. How can we advance the dialogue with industry and business? Do we still need to justify the environmental factor or have we passed that stage?

For the purpose of today's discussion, a basic question that needs to be answered is: 'why should British industry and investment be concerned about the protection of the environment in the developing countries?' You are of course aware that there is much talk now of reindustrialization, of moving to a post-industrial economy, with industries such as microelectronics and biotechnology coming to the fore. With the rapid growth of the new industries and the replacement of some raw materials with synthetics, I am aware we can no longer look upon the UK economy simply as an exporter of manufactured goods and an importer of primary products. The fact that the UK's trade figures showed this year that, for the first time since the Industrial Revolution, it had become a net importer of finished goods underlines the point.

However, it must not be forgotten that now, and for the foreseeable future, the UK will stay in part reliant upon the developing world for food, energy and industrial raw materials. This country continues to rely for its industrial processes on a wide range of natural products, such as dyes, resins, pectins, tannins, fats, waxes, pyrethum and other natural pesticides. For the most part they come from the Third World, as do a number of foodstuffs, such as sugar, coffee and tea. In turn, the UK requires markets for its output of finished goods and services. According to OECD, 30% of the UK's visible exports still go to the developing world. I am sure you are more conscious than I that this nation is part of interdependent natural and economic systems.

Knowledge of the environment

Last year UNEP published a major scientific report on the changing condition of the world environment since 1972. It found that our knowledge of the environment in the global South is sketchy at best. But we do know enough to say that most less developed nations are facing an unprecedented environmental crisis. Deforestation, soil erosion and desertification, misuse of water resources and even the familiar threats in the industrialized countries of water and air pollution are undermining the human and bioproductivity of these nations.

Some aspects of poor world environmental destruction pose obvious dangers to the economies of Western nations. Take, for example, the issue of genetic diversity. The tropical forests, savannah and arid regions of the developing world are the world's genetic storehouses. (One small island in Panama contains as many plant species as are found in the whole of the British Isles.) Agriculture, pharmaceutical and even the new biotechnology industries are, to varying degrees, dependent on fresh infusions of genetic material from these wild factories. But virtually everywhere they are under threat; at a conservative estimate some 25000 plant species are known to be threatened with extinction. Ten percent or more of all species on earth could be extinguished over the next two decades. Many threatened species have not even been tested for their potential industrial utility and this all relates to the species we know. Those we do not know about are many times more in number.

For the most part, ecological connections are difficult to make. Through television and newspapers the public in the West has become familiar with and even inured to images of destitution, civil conflict and bloodshed in the Third World. But I would put it to you that few people in the UK appreciate how widespread and how closely related these disasters are to the misuse of natural resources and environmental destruction. Take, for example, the appalling landslides and flooding which have claimed the lives of hundreds in Nepal in recent weeks. The loss of life was reported in *The Sunday Times*, but no effort was made to explain that the cause of the disaster was deforestation on the hillsides.

The ecological link

Before us looms the spectre of conflict over threatened resources, over what might be termed the security of nature's supply. I would like to quote a section on the ecological link from the recently published Conservation and Development Programme for the UK:

This [ecological link], generally, is more uncertain and less visible than economic and political threats, but it is no less real. The ecological linkage produces direct effects upon supply . . . We may feel the impact first through political instability and social strife overseas. In the Third World, famine and drought have, in the last decade, started or intensified political conflicts, for example between Ethiopia and

Somalia in the 1970s and in India and Bangladesh. Closer to home we see not merely economic loss through ecological damage, for example in the loss of fisheries through over-fishing, but we also experience the political tensions that this can bring.

With a projected doubling of withdrawals from the world's water supply between now and the year 2000, we should be alert to the danger of conflict over how nations utilize this basic resource. Of the world's major river basins, 148 are shared by two countries and 52 by between two and 10 nations. We have already seen how dam-building, pollution, siltation and unsustainable withdrawals have angered nations downstream. On a world scale, such tensions feed global instability, which encourages big power rivalries. This in turn fuels the arms race. The process loops back into the other linkages, intensifying global interdependence. The prospect of climatic warming, which may well redistribute economic advantage in the production of food, the most strategic material of all, suggests that the ecological link could become the dominant connection in the interdependence of the coming century.

The need for reform

It is my conviction that we are quickly running out of time to reform a global economic order that takes so little account of the sustainable and equitable use of renewable resources. As an international civil servant it is not my role to pronounce on the relative merits of capitalist, planned or any other politico-economic system. Our job in the UN is to work as best we can within the systems devised by sovereign states. But common sense demands that the economic order dating to Bretton Woods is in peril. The Brandt Report laid out how this system could be reformed but little has been done to implement its prescriptions.

Unfortunately, the case for sustainable development is as yet unbalanced. In the less developed countries it rests overwhelmingly on what has gone wrong rather than what has gone right. The catalogue of woes is lengthy: the energy potential of hydro-electric power schemes reduced in some instances by half through siltation; uncontrolled urban development swallowing up prime farming land; a levelling out since 1970 of the world's fish catch through overexploitation; short-sighted logging turning former exporters of hardwoods in West Africa and South-east Asia into net importers; agricultural projects from semi-arid to tropical, moist forest regions collapsing because no account was taken of the soil's carrying capacity; foreign-financed cash crop schemes driving villagers onto marginal lands. These are but a sample of the threat environmentally destructive development poses to the very fabric of the newly independent nation states. All too often these projects have been introduced with little or no effort to anticipate the environmental consequences. Where environmental impact assessments have been made, the trend has been to ignore them when construction begins.

Few aspects of Third World development typify this malaise better than irrigation. To hard-pressed decision makers in developing nations, irrigation appears to offer an escape from the twin pincers of dwindling foreign exchange reserves and food supply barely able to keep pace with population increase. So again and again we see the triumph of hope over experience. Soon the schemes become plagued by salinization and waterlogging. According to the Club du Sahel, as fast as new land goes under irrigation in the Sahelian region, existing areas go out of production. This is a pattern repeated with depressing regularity throughout the developing world.

The pity of it is that we have the know-how and the manpower to prevent irrigation and other kinds of development from going wrong. Benjamin Franklin's two centuries-old dictum, 'an ounce of prevention is worth a pound of cure', is as relevant today as it was then.

I must concede that five years ago I was delivering much the same message. There are those, even among my colleagues at UNEP, whose counsel is to play down the crisis and concentrate on the opportunities and challenges. Quite rightly, they fear that the response to an increasingly difficult world plagued by overpopulation and fewer and fewer resources *per capita* will be one of retrenchment. This is an eventuality that must be avoided at all costs.

Balancing opportunity and threat

I believe that when talking to the practitioners of development we need to balance opportunity and threat. One of the opportunities is presented by a new receptiveness among Third World decision makers to the environment factor, a development which has been all but ignored in the Western media which continues to report the *coup d'etats* and the disasters, but little else.

Though the need for foreign exchange is still desperate, there is a new willingness to balance long-term productivity and the need to meet people's basic needs with the requirement for hard currency. Certainly, in UNEP we are being overwhelmed by requests for advice and assistance from the developing nations on how to go about building up environmental management self-reliance. This is a process which, for the most part, does not demand massive infusions of capital, sometimes meaning little more than redeploying manpower, know-how and resources.

New business opportunities are being opened up by this gathering interest in environmentally-sound development. One legacy the UK enjoys from its traditionally close trading links with Commonwealth partners, and also from its pioneering of environmental sciences, is a wealth of environmental expertise. It is present not only in firms but also in development institutions, government and the universities. We need look no further than the UK Conservation Strategy, which was a collective effort by British NGOs, as evidence of the depth of this country's environmental understanding.

The environment is becoming big business, and the UK has a head start. UNEP's studies have revealed that some firms, by building resource conservation into plant development, have boosted profits, sometimes by as much as 30%. By vigorously applying a resource conservation and recycling programme, the giant 3M Company of the USA produced total savings of US$97 million between 1975 and 1981. In order to make an equivalent amount of earnings at 3M's traditional profit level would require additional sales in the order of US$300 million over that period.

On a smaller scale, a distillery in Scotland, faced with an ultimatum on pollution control, decided to increase, by evaporation, the proportion of solid matter in the spent 'wash' from its stills from 2% to 40%. Processing charges amounted to some £300 000 per annum and capital charges to a further £200 000. On the other hand, revenue from the sales of dried wash as cattle feed amounted to over £1 million. The net profit was thus over £500 000 per annum, and the pollution problems, which triggered the whole exercise, were largely eliminated in the process.

Although the science of environmental cost–benefit analysis is in its early stages and many imponderables remain, the evidence appears overwhelming that a company can use the environmental factor to boost both profits and growth. Recent estimates have put the value of the world market in pollution control and other types of environmental goods and services at some US$100 billion. Here and on the continent new firms have been set up to meet the demand, and many of the older established companies such as Exxon, Dow, Krupp, ICI, Shell and Philips have now created new divisions to exploit this new market. In UNEP we have been hearing of many new initiatives on recycling wastes from countries as diverse as PR China, Brazil and South Korea.

There are signs, too, of a developing appreciation of the nature of the ecological connection I mentioned earlier. Governments in UNEP's Governing Council have demonstrated their conviction that we need – in terms of policy implications – to develop our ideas on the ecological link between North and South. It will be the subject of UNEP's 1984 State of the Environment Report and will bulk large in the report of a Special Commission on environmental perspectives which UNEP is helping to establish. The Commission's brief will be to set environmental goals for the year 2000 and beyond.

However, the case for environmental protection does not and can never rest entirely on economics and mutual self-interest. Numerous public opinion polls, here and abroad, testify to peoples' concern for the maintenance of environmental quality. For a mix of aesthetic, moral and spiritual reasons they care not only about their local surroundings but also about the fate of tropical forests, pollution of the atmosphere, seas and so forth. Industry and commerce, even for reasons of public relations alone, should be concerned about the environmental factor. A recent MORI Poll found that more than 58% of the UK public would countenance a penny in the £ increase in income tax to finance conservation.

For practical and ethical reasons industry too should see to it that uniform environmental standards are applied. The UK has a 12.2% share of the world pesticide export market, with nearly half destined for the developing world, but in 1981 an Oxfam survey found that British industry was exporting at least a dozen types banned in industrialized countries as being too hazardous. It is a double standard that can boomerang – a US government survey found recently that a third of all imported coffee beans were contaminated with pesticide banned at home.

I expect some of you, on your flights overseas, have seen the advertising sequences which precede the main inflight film. I saw one recently in which a major private investment bank boasted – over scenes of tropical timber being felled and caterpillar tractors bulldozing a new road through the forest – of taming the limitless jungle to bring prosperity to all; suddenly I was back in the 1960s with progress couched exclusively in terms of exploitation. If there were environmental safeguards they were not considered important enough to merit a mention.

Thousands of such project investment decisions are taken each year in the City of London. In UNEP, we realize that until we make an impact on the private investment banks in the City of London and other financial centres of the world we will fail to make adequate progress in halting resource destruction in the developing countries. Earlier I referred to the danger posed by retrenchment. There are disturbing signs that the private investment banks are becoming increasingly reluctant to invest in the less developed nations. The challenge facing us in the environment and development community is to follow-up our progress with the multilateral and bilateral institutions by convincing private business that the gathering support for sustainable development in the Third World is opening up a new field for profitable investment.

The role and work of UNEP

I should like to end my address by talking very briefly about UNEP – our role and how, in partnership with industry, we can offer each other assistance. UNEP, with headquarters in Nairobi, was the first UN body to be established in the developing world. Our brief is a particularly difficult and challenging one; unlike, say, FAO or WHO we do not, by and large, execute projects in the field – our assigned role is to persuade or 'catalyse' UN agencies and the other practitioners of development to take the environment into account at all levels of decision making. This is a monumental task when you bear in mind that we have, world-wide, a professional staff of about 180. By way of comparison, the Greater London Council's Architects Department employs 10 times as many professionals. As governments attending our 10 years commemorative session last year recognized, we have made an impact way out of proportion with the small budget and staff we have available. That role will be enhanced as, in partnership with the rest of the UN system, we set about the work of applying the newly launched System Wide Medium Term Environment Programme (SWMTEP). This programme, the first of its kind, will ensure that the UN makes a more concerted and coordinated response to the environmental challenges of the 1980s.

UNEP also operates a small Industry and Environment Office in Paris. Since 1975 we have been expanding our consultative dialogue with business and industry. One product has been a steady stream of balanced reviews and guidelines on environmental management for specific industrial sectors. Any ideas on how we can cooperate on the promotion of environmentally sound industrial practices and on the transfer of the wealth of industrial experience from the industrialized countries to the developing world would be welcome. There would seem to be plenty of scope for practical partnerships through, for example, proper mechanisms for dissemination of information, staff secondments, education and training ventures and scholarships.

UNEP also operates a fully computerized service on environmental information sources called INFOTERRA. INFOTERRA operates along the lines of the telephone directory's yellow pages in that it does not provide the information directly but will tell the user which of its 10 000 or more sources will provide the data being sought. There is a world-wide network of 119 INFOTERRA national focal points, including one in the UK.

World conference on industry and environment

A vital part of our outreach strategy will be the convening next year of a world conference on industry and the environment. Since early last year we have been in close contact with leaders from a cross-section of international business. The reactions from all quarters, including our Governing Council, have been extremely positive. We have set up a committee of conveners made up of prominent figures from industry and business, government, international organizations and the non-governmental community. Industry itself will be footing most of the bill for the conference. The conference will deal with policy rather than the techniques of environmental management. Three main objectives have been recommended by UNEP's Governing Council:

1. to consider ways in which industry might contribute more fully and effectively to environmentally sound development;

2. to promote a broad exchange of information and experience on technologies and institutional approaches to industrial development, including industrial pollution control, recycling of wastes and low- and non-waste technologies;
3. to consider ways in which the extensive technical resources of industry in both developed and developing countries can be mobilized and applied more effectively to environmental management, including the identification of constraints and opportunities.

Conclusion

In asking me to address this seminar on the environmental factors in Third World development, the organizers presented me with a difficult task, notably how to compress such a broad and diverse subject into something meaningful for the ensuing discussions. I hope you appreciate that the issues I discussed were by way of being introductory in nature. Important aspects such as cost–benefit analysis and environmental impact assessment were only touched upon, while other relevant aspects, such as the mechanisms for integrating environment into development planning, were a casualty of 'cut and snip'.

14

Environmental management in the Gulf

Lecture to the University of Qatar

Qatar, November 1983

The sea has nurtured the civilizations of the Gulf region for many centuries. Long before the great voyages of the Europeans, Arab traders had established trade routes to East Africa and beyond. The dhow is as much a symbol of Arab civilization as the Haj or the noble lifestyle of the Bedouin.

The Arab nations of the Gulf continue to depend on the sea: not only for trade, but for recreation, for fishing, oil transportation and now even for drinking water. But the coastal zone, which has for so long supported the people of Qatar and its neighbours, is under threat from the hyperdevelopment that has come to this region. It lies within the power of the governments and people of the Gulf area to remove that threat; properly managed, the shallow coastal area of the Gulf can continue to sustain your civilization during this time of rapid economic transformation.

A 16th century English philosopher once wrote that 'Nature, to be commanded, must be obeyed'. The Industrial Revolution has given mankind the illusion that he is free from the constraints of nature; that science and technology can be used to subdue the natural systems which support all life on earth. Notwithstanding some positive developments, we have to say that technology is still most often used in the service of a development process which destroys more that it develops. Our job in UNEP is to persuade the practitioners of development of the vital need to stand that process on its head.

Marine pollution, tarred beaches and declining fish catches are merely warnings that we are pursuing the wrong kind of development – one that can be reversed if we change our sense of purpose. That purpose should be to institute economic development which, at all levels, takes environmental considerations into account. We call this 'sustainable' or 'environmentally sound development'. Soil erosion, deforestation, air pollution and so forth are further symptoms that we have still to inject an environmental purpose into development. The global community is already paying the price in the form of undermined food production, contaminated rivers, narrowing genetic diversity, acid rain and a resurgence of environmental diseases.

Little environmental destruction that has taken place has been irreversible. We have sufficient resources and know-how to provide for the needs of our growing world population. But those resources must be used wisely and equitably. Helping to bring this about is UNEP's main preoccupation.

My intention is to talk briefly about environment in the wider context of development, before moving on to discuss in detail coastal management. You would agree that to do otherwise would be to present an unbalanced view.

Environmental protection

One of the main constraints on the work of UNEP and, indeed, of the environment movement as a whole, is the misconception that environmental protection is a minority pursuit. Today there is still a widely held belief that environmental management is an activity that can somehow be carried on in isolation. The extent to which our environment is interconnected, with neglect in one quarter frustrating positive action in another, is one of the main lessons of the past 10 years. In the 1980s, the challenge is to get industrialists, planners, bankers, farmers, government and UN officials and the other participants in the development process to agree common plans and programmes that use resources fairly and sustainably.

However, the case for environmental protection does not and can never rest entirely on economics and mutual self-interest. Numerous public opinion polls testify to peoples' concern for the maintenance of environmental quality. For a mix of aesthetic, moral and spiritual reasons, they care not only about their local surroundings but also about the fate of tropical forests, pollution of the atmosphere, seas and so forth. The time has already passed when we needed to look upon clean air, water, topsoils, wild plants and forest cover as resources every bit as precious as, say, coal, oil or mineral ores. We have traditionally taken these self-regenerating or 'renewable' resources for granted. However, these resources are renewable only if conserved, finite if not. Through a combination of avarice, irresponsibility and ignorance we have, in effect, been mining these sustainable resources, and we have seen the consequence in the form of resources depletion.

Avarice, by which I mean overconsumption and the pursuit of wasteful lifestyles so obvious in large segments of societies in developed countries, is not a major cause of environmental degradation within the developing countries relative to poverty. Most of the poor nations of the Third World are locked into a cycle of resource destruction and deprivation. Handicapped by population increase, unfair terms of trade and initial poverty, the governments of these countries have little scope for preventing their poor, who live on the margins of existence, from destroying the very resources on which their future livelihood depends. This process was accurately described recently as one of 'self-immolation'.

But against this rather bleak scenario, I can report that the environmental factor in world development has begun to play an increasingly important role; we have made rapid headway among the decision makers who work in government, the UN agencies and the development assistance agencies. Few, if any, in those circles would put a contrary case that development which takes no account of the environment is worth pursuing.

Perhaps the most obvious sign of this transformation is a change in the environmental vocabulary: the emphasis is no longer on 'small being beautfiul' or on 'limits to growth', instead we talk exclusively in the terms I have mentioned such as sustainable development and rational resource management.

These are not new catchwords, but significantly none have come along to replace them. The overwhelming concern of the responsible environment movement nowadays is to see these concepts put to work. The world now desperately needs skilled manpower and money to move the ideas of sustainable development into meaningful action.

The view now generally prevails that the overall environmental objectives are:

● to raise indigenous environmental or natural resource management capability;

- to build upon existing experience in the North so as to ensure that lessons are learned from their mistakes of the past;
- to ensure that environmental planning is not left out in development planning;
- to gather sufficient hard data of an environmental kind (taxonomic, ecological, geological) to enable sound development planning to take place;
- to inform the public of what is at stake;
- to concentrate on systems particularly at risk, be it deserts, watersheds, moist forest or areas of rapid urban expansion.

Sustainable development

In conventional terms, the concept of sustainable development encompasses:

1. help for the poorest because they have no option but to destroy their environment;
2. the idea of self-reliant development, within national or local boundaries and within natural resource constraints;
3. the idea of cost-effective development, but often on different time scales to traditional economic criteria; that is to say development should not degrade environmental quality, nor should it reduce productivity in the long run;
4. the great issues of health control, appropriate technologies, food self-reliance, clean water and shelter for all;
5. the notion that people-centred initiatives are needed; human beings are the resource in the concept.

The multilateral development assistance agencies, which spend between themselves more than US$15 billion on new projects each year, have radically altered their lending policies in favour of the environmental factor. The turning point was the signing in 1980 of a declaration which formally committed the Arab Bank for African Development, the World Bank, UNDP and the other major lending and development financing institutions to environmentally sound development. The purpose of the Declaration was perhaps best summed up by World Bank President Clausen when he said in a major public statement that 'sustainable development and wise conservation are, in the end, mutually reinforcing and absolutely inseparable goods'. Since then, several country-to-country aid agencies, including those of the USA, FR Germany and Sweden, have made a similar commitment.

But the amount disbursed by the multilateral and bilateral organizations pales into insignificance when compared to the investments industry itself and the private financing institutions make each year. I have no mandate to speak on behalf of our partners in the environment and development community, but I believe they would not disagree with our view in UNEP that, in comparison, our impact on private business and industry has been small indeed. A major priority for UNEP in the next few years will be to increase the dialogue with industry and business. The starting point will be a world conference on industry and environmental management next year.

Though the rapid growth of the new industries and the replacement of some raw materials with synthetics is changing the pattern of global trade, we can say with confidence that for the foreseeable future the oil exporting Arab countries and the Western nations will stay in part reliant upon the other parts of the developing world for food, energy and industrial raw materials.

Some aspects of poor world environmental destruction pose obvious dangers to the economies of Gulf nations. Take, for example, the issue of genetic diversity. The tropical forests, savannah and arid regions of the developing world are the world's genetic storehouses. Agriculture, pharmaceutical and even the new biotechnology industries are, to varying degrees, dependent on fresh infusions of genetic material from these wild factors. But virtually everywhere they are under threat; at a conservative estimate, some 25 000 plant species are known to be threatened with extinction. More than 10% of all species on earth could be extinguished over the next two decades. Many threatened species have not even been tested for their potential industrial utility, and this all relates only to species we know. Those we do not know about are many times more.

Ecological connections are difficult to make. Through television and newspapers, the public has become familiar, even inured, to images of destruction, civil conflict and bloodshed. But I would put it to you that few people appreciate how widespread and how closely related these disasters are to the misuse of natural resources and environmental destruction.

The danger of conflict over threatened resources

Before us looms the spectre of conflict over threatened resources; over what might be termed the security of nature's supply. The ecological link is generally more uncertain and less visible than economic and political threats, but it is no less real. The linkage produces direct effects on supply. We may feel the impact first through political instability and social strife overseas. In the Third World, famine and drought have, in the last decade, started or intensified political conflicts, for example, between Ethiopia and Somalia in the 1970s and in India and Bangladesh.

With a projected doubling of withdrawals from the world's water supply between now and the year 2000, we should be alert to the danger of conflict over how nations utilize this basic resource. Of the world's major river basins, 148 are shared by two countries and 52 by between two and 10 nations. Underground water aquifers are also shared by two or more states in many cases. In arid and semi-arid regions almost everywhere these supplies are coming under increasing pressure. Overusage in some areas has even led to what amounts to the mining of fossil water supplies. With the advances in well-drilling and pumping, the Gulf region's six significant fields of fresh groundwater are coming under increasing pressure. There is evidence of environmental damage as a result of overpumping, subsidence due to compaction following pumping, intrusion of saline water and contamination through polluted surface water recharge. These changes can be irreversible, or can be corrected only over very long periods of time.

In the Gulf region, a recent UN inter-agency field misson found that freshwater extraction is reaching a serious stage. The report of the mission found that reserves are being withdrawn faster than the natural replenishment rate. Countries in the region, both individually and collectively, still have time to head off the problem before it becomes critical. As in Western Australia, it may become necessary to impose strict regulations on withdrawals. Certainly the situation calls for careful assessment of the status of reserves to determine their sustainable capacity and continued monitoring. Advances in engineering are also needed to help conserve supplies by water recycling and controlled mixing with brackish groundwater and desalinated water.

We have already seen how dam-building, pollution, siltation and unsustainable withdrawals have angered neighbouring nations. On a world scale, such tensions feed global instability, which encourages big power rivalries. This in turn fuels the arms race. The process loops back into the other linkages, intensifying global interdependence. The prospects of climatic warming, which may well redistribute economic advantage in the production of food, the most strategic material of all, suggests that the ecological link could become the dominant connection in the interdependence of the coming century.

The need for reform

It is my conviction that we are quickly running out of time to reform a global economic order that takes so little account of the sustainable and equitable use of renewable resources. One problem we have to overcome is the uncritical acceptance of alien development patterns. We have been forgetting our traditional knowledge of, and respect for, the environment. It is significant that in Iraq, when a major multimillion irrigation scheme went wrong, the local people averted total disaster by switching to ancient drainage methods.

For thousands of years, in some cases, villagers and townspeople have used local resources prudently so that they have been handed over intact from one generation to the next; often this has been accomplished unconsciously as they observed religious, cultural and other practices finely honed to their environment. In terms of material possessions, these people were poor, but in their respect for what nowadays we refer to as the balance between people, resources, environment and development they were rich indeed. The renewed interest in the 'afflaj' system of drainage and the attempts to revive modern versions of the Bedouin Hema system of rangeland management are encouraging signs that decision makers in the Arab nations are beginning to rediscover their environmental heritage.

Business opportunities

New business opportunities are being opened up by the gathering interest in environmentally sound development. In the industrialized countries, the environment is becoming big business. UNEP's studies have revealed that some firms, by building resource conservation into plant development, have boosted profits, sometimes by as much as 30%. By vigorously applying a resource conservation and recycling programme, the giant 3M company of the USA produced savings of US$97 million between 1975 and 1981. In order to make an equivalent amount of earnings at 3M's traditional profit level would require additional sales in the order of US$300 million over that period.

Although the science of environmental cost–benefit analysis is in its early stages and many imponderables remain, the evidence appears overwhelming that a company can use the environmental factor to boost both profits and growth. Recent estimates have put the value of the world market in pollution control and other types of environmental goods and services at some US$100 billion. New firms have been set up to meet the demand, and many of the older established companies, such as Exxon, Dow, Krupp, ICI, Shell and Philips, have now created new divisions to exploit this new market.

There are signs, too, of a developing appreciation of the nature of the ecological connection I mentioned earlier. Governments in UNEP's Governing Council have demonstrated their conviction that we need, in terms of policy implications, to develop our ideas on the ecological link between North and South. It will be the subject of UNEP's 1984 State of the Environment Report and will bulk large in the report of a Special Commission on environmental perspectives UNEP is helping to establish. The Commission's brief will be to set environmental goals for the year 2000 and beyond.

Preserving resources of the coastal environment

Helping coastal nations to keep their shared waters productive is a vital part of that strategy. So many of us depend on their continued health. Since the dawn of civilization, we have been an atavistic people. Seven out of 10 people around the globe live within 80 kilometres of the coast. Almost half the world's cities with a population of over one million are sited in and around the tidal river mouths. One reason for their popularity is obvious: coastal zones provide all but 10% of the world's fishing catch. In many countries, fish is the major source of animal protein, accounting for 55% in Asia, for example.

The mudflats, coral reefs, shallows, estuaries, caves, mangroves and beaches are the environment for most forms of marine life. Coastal zones are where we find a large majority of the 20 000 known varieties of fish, the 30 000 species of mollusc and almost all the crustaceans. Apart from ourselves, many birds and animals rely on the sea's coastal harvest.

Coastal zones, however, are precisely where human beings put most pressure on the marine environment. We use the coastal areas for our settlements, as our food store, as a playground and as our rubbish dump. It has been calculated that something less than 10% of all the material entering coastal waters reaches the open ocean. The rest remains in the coastal sediment. Sludge from domestic sewers has almost completely stifled once productive and valuable shellfish beds in the waters near several North American cities. In Indonesia, dramatic declines in shrimp harvests have been recorded in areas cleared of mangrove.

Balancing development with preservation of the coastal environment

The challenge is to strike a balance; one that will preserve the coastal web of life, while satisfying the legitimate need to develop.

But what should we do about the wastes from our factories, our homes, our ships? They have to go somewhere. How poisonous are they? We have to consider our neighbours as well as ourselves: DDT traces have been found far from anywhere the pesticide has been used, in Antarctic penguins and Arctic seals. As yet, there are no reports of any harmful effects on human beings. By contrast, scores of people in the Japanese village of Minamata died when they ate fish contaminated by the wastes from a nearby industrial plant.

How fast can we develop without endangering the environment, the life around ourselves? It is not just a question of industrial development. The world population today is about six times higher than it was 200 years ago on the eve of the Industrial Revolution. All these people need to be housed, fed, kept healthy if possible, and employed.

The coastal ecosystems intermesh with terrestrial and open ocean systems to form a complex whole. The tides and currents that make coastal waters the ocean's most biologically productive regions also render them exceptionally vulnerable to pollution. Damage one part of the system and a chain reaction will start, thus deforestation in watersheds located hundred of kilometres inland can result in silt-choked harbours and coral reefs.

To be sure that we are living in balance with the environment, we first need to know how bad the pollution is: how much we produce and how toxic are its effects. We need to learn how to manage our environment to make sustained use of its resources and how best to legislate to protect these resources. Any rational conservation programme must provide for both scientific assessment of the problems and environmental management; and this can be achieved effectively only through international cooperation.

The need for regional cooperation to protect shared resources

It cannot be stated too often that environmental problems rarely affect one nation alone, particularly in coastal areas and the marine environment. Each country's pollution, whether from land or offshore, can degrade the environment of neighbouring states. Within a region, fishing grounds are normally shared by several nations.

Most of the seas' environmental problems show up in coastal waters and are often specific to a particular region. Therefore, in the view of UNEP planners, regional cooperation is needed to protect the coastal environment and the sea's resources, whether the problems are oil spills, pollution from land or destruction of animal habitats under pressure from human settlements.

The problems of pollution from sources such as industrial waste, municipal sewage and agricultural run-off starts on land. Together these sources account for 80% of all pollution reaching coastal waters. In the present unequal state of world development, it would appear impossible to unite all countries in a global programme to control such pollution. However, within individual regions it has proved feasible. Enlightened self-interest has been the spur; governments can see clearly their common interest in holding down such hazards to the marine environment when they all share the same water resources.

UNEP's Regional Seas Programme

UNEP's genesis was the 1972 Conference on the Human Environment, organized in Stockholm by the United Nations. The Stockholm Conference underlined 'the vital importance' for humanity of the seas and all the living organisms which the oceans support. In its Action Plan, recommendation 92 said governments should take 'effective national measures for the control of all significant sources of marine pollution, including land based sources, and concert and coordinate their actions regionally and where appropriate on a wider international basis'. UNEP's Governing Council endorsed the regional approach to controlling marine pollution before UNEP started its Regional Seas Programme in 1974. We decided at first to concentrate on four regions: the Mediterranean, the Caribbean, West Africa and the Gulf region.

There are now 10 Regional Seas Programmes, involving 120 states, 14 United Nations agencies and 12 other international organizations, in UNEP's efforts to help protect the marine environment in these regions. A Regional Seas Programme Activity Centre, set up in Geneva in 1977, coordinates the work carried out under the Programme. For each region, UNEP has adopted a basic strategy aimed at tackling the causes as well as the consequences of environmental damage in coastal areas. This strategy encompasses:

1. an action plan setting out activities for scientific research and cooperation, including assessment and management;
2. a legally binding Convention embodying general commitments;
3. technical and specific protocols to deal with individual issues, such as dumping, cooperation in pollution emergencies, land-based pollution sources, and conservation;
4. financial and institutional arrangements that provide the back-up for the other three parts of the strategy.

Coordinating all this activity can take several years. The Kuwait Action Plan took more than four years before it was adopted.

UNEP acts only at the invitation of governments in putting together a regional programme and involves governments from the beginning in formulating an action plan. Once the plan is adopted, national institutions are nominated by their governments to implement the programme. UNEP and other international organizations provide the seed money for the programme. As a programme develops, governments in the various regions take over financial and, whenever possible, administrative responsibilities. The programme for the Kuwait Region is, to all intents and purposes, already independent of UNEP.

Regional Seas Action Plan

In the four interdependent areas of activity in the Regional Seas Action Plan – assessment, environmental management, legislation and support measures – first priority goes to assessment and evaluation of the sources, amounts and effects of pollutants, the state of living and material resources, and analysis of development practices which make a direct or indirect impact on the environment. The results are designed to help national policy makers in managing their natural resources effectively and sustainably; development without destruction is the guiding principle.

The second area of activity, environmental management, aims to help managers improve their ability to make decisions on their own, and to develop integrated plans for coastal area development. The third, legislative section includes the regional convention and protocols. The plan can also help governments bring their national legislation on the environment and natural resources into line with regional partners. Finally, support measures are required to enable all countries in the region to take part fully in an action plan. Scientists and policy makers may require technical assistance. Marine laboratories may need help to install sophisticated equipment. Assistance may also be needed to ensure information gathered is standardized and comparable.

The overall aim of our Regional Seas Programme is to enable national institutions eventually to take over responsibility for even the most technically demanding aspects of an action plan. At the same time, common elements are closely coordinated and will eventually lead to a global ocean programme; so it is necessary to make sure that the data collated in this region, for example, are also internationally useful.

The Mediterranean

The forerunner of this global Regional Seas strategy was the Mediterranean. A miniature ocean, bordered by 120 cities with a coastal population of at least 100 million, the virtually enclosed waters of the Mediterranean Sea have been the well beaten crossroads of European, Asian and African civilizations for at least 4000 years of recorded history. The pollution of its waters led many to fear it was at risk of dying. Once a symbol of the sea's beneficial impact on man, it became a symbol of man's destructive impact on the sea itself.

Against a background of gathering international concern for the welfare of the Mediterranean, UNEP helped the countries of the region to adopt, in February 1975, an action plan for the protection and development of their common sea. What happened in the Mediterranean had a significance beyond its boundaries. As the programme evolved, we were able to broaden and deepen our understanding of the role of environment in the overall development process. And just as important, we saw that nations, developed and developing, Islamic and Christian, were prepared to put aside their political differences to cooperate in an endeavour to protect their shared environment. In UNEP we like to think that the goodwill generated has spilled into other spheres.

The Gulf region

The nations of the Gulf region face similar perils and opportunities, though the origins of both are somewhat different. For the countries of the Red Sea, Gulf of Aden and the Gulf, two centuries of industrialization have come all at once. Oil wealth has brought both the benefits and problems of fast economic growth to two of the world's most fragile environments.

The waters of this region are shallow and virtually landlocked. Shore waters, less than 10 metres deep, stretch for many kilometres off the 1200 kilometre-long Gulf. They receive almost no rain and hardly any fresh water except through the Shatt-al-Arab waterway. Sea water comes through the Straits of Hormuz, but much is lost through evaporation. This means that pollution is not flushed away easily. Qatar, in short, shares one of the world's most brittle and endangered environments.

Qatar and seven of the other fastest developing countries of the world share the shores of the Gulf. Migrants crowd into the towns, some of which are doubling in size every four years, and in many countries nearly everyone lives on the coast. Existing sewage and waste disposal systems cannot cope and it will be some years before all the states of the region have modernized systems in operation. In some places, 75% of the sewage pours untreated into the coastal waters. Qatar and Kuwait – UNEP is encouraged to note – will be the first nations to install an adequate system.

Some 20 major industrial centres are being developed, again almost all on the coast. Investment has been estimated at US$20 – US$30 million per kilometre of coastline. But no states, so far, have well established integrated programmes of pollution control, although some individual industries both in Qatar and in neighbouring countries have set their own high standards for pollution control.

Controlling oil pollution in the Gulf

As for oil, almost two-thirds of all the petroleum carried by ships is exported from this region, and the level of oil pollution in its waters has been calculated at 3.1% of the global total – 47 times the average for a marine environment of its size. One recent survey found that tankers carrying petroleum from Gulf terminals dump close to one million tonnes of oil into the sea each year, through discharges of ballast water from tankers containing residues. One obvious manifestation has been the tarring of Oman's superb beaches on the Arabian Sea.

Taking an overall picture of the state of the environment in the Gulf, it must be said that the quality has deteriorated, is deteriorating and will continue to deteriorate unless the resources and know-how the countries of this region possess are mobilized.

Already, the per capita loads of solid wastes produced by the industrial plants in the Kuwait Action Plan Region are higher than the average of the developed countries. In two Gulf countries they are more than double the specific amounts estimated for the developed western countries. Industrial wastes are mainly the oily and toxic sludges generated by crude oil and petroleum product storage in oil export terminals and refineries. Though, as I have said, there are encouraging signs that pollution from sewage and other domestic sources is coming under control, all the current trends indicate that pollution from industrial sources will continue to increase.

Tough new regulations are needed to control the principal sources of oil pollution, namely offshore drilling and transport. Discounting ballast pollution and major oil spillage incidents, at current rates we forecast that more than 1.5 million metric tons of oil will pollute the Gulf during this decade. When voluntary restraints are not forthcoming, industry and the oil companies, in particular, should be made to put their house in order.

Air polluton in the Gulf

An increasing problem is air pollution. A recent inventory found air pollution to be on the increase in the newly developing coastal areas. The principal pollutants are carbon monoxide from motor cars, and particulates from industrial processes which also produce half of the hydrocarbons and all of the significant ammonia loads.

Natural gas flaring is the single most important polluter as far as sulphur dioxide is concerned, putting into the air over 2 million tons of SO_2 each year, about 65% of the total. The move towards collecting the natural gas now being flared to burn for energy and for use in petrochemical manufacture in the new industrial zones will concentrate SO_2 releases in areas closer to where most people live. Strict sulphur recovery technologies need to be installed now before mistakes are made.

Enforceable regulations and penal codes are also needed to keep industrial pollution to a minimum. As I stated earlier, companies in the industrialized countries are boosting profits by building in resource conservation and pollutant recycling; the same could be achieved in this region.

The Red Sea and Gulf of Aden

On the other side of the Arabian Peninsula, the Red Sea and Gulf of Aden region is just feeling the effects of new wealth. Today oil is the fuel for development, but the promise for tomorrow is that onshore and offshore mineral mining will provide even more earnings. Though the Red Sea region is relatively free of pollution and unaffected by population pressures, cities and industries are growing fast, along with oil exploitaton and shipping.

What makes the Red Sea and Gulf of Aden of particular interest to geologists and biologists is that it seems to be a new ocean in the first stages of formation. It has some of the world's most northerly coral reefs, with numerous endemic species of animals and plants.

Cooperation in the Gulf

The political tensions which are present between some countries of the two regions are probably greater than those in the Mediterranean. But to the enduring credit of the Gulf countries, we have once again seen a preparedness to take the environment out of the political cockpit. They have already recorded some unparalleled achievements in environmental cooperation. Despite the pressures for fast development, the 1982 Red Sea and Gulf of Aden Action Plan is unique in the Regional Seas Programme for putting its main emphasis on conservation. The Kuwait Region's Convention, adopted in 1978, entered into force in just two years, and the implementation of the region's Action Plan has never lacked for money – always the first test of any international commitment.

The Kuwait Action Plan covers cooperation between Iran and the seven Arab States of the Arabian Peninsula on oil pollution, industrial wastes, sewage, fisheries resources, and the environmental impact of coastal engineering and mining. Projects range from public health, fish farming, marine parks, port pollution, freshwater management, to the development of a mathematical model describing the physical oceanography of the Gulf. These are achievements for which the countries of this region can take an immense pride. In UNEP, we are confident that, in a renewed spirit of cooperation, action will be taken to stop the damaging oil spill in the Gulf.

I understand three of the seven wells which are spilling oil have been capped. This will have reduced the flow of oil which was estimated to be running at between two and 10 000 barrels a day. Though there is evidence from this region and elsewhere that oil pollution causes considerably less havoc than was once thought, this is no cause for complacency. The threat to the natural ecosystems, fisheries, coral reefs and to industrial installations such as desalinization plants is very high. Perhaps the greatest danger is the cover the continuing oil spillage provides for unscrupulous polluters. In this, as in all aspects of action to improve the Gulf region's environmental protection, UNEP stands ready to assist.

Harsh experience has taught us that such incidents are certain to occur. The key to an effective oil spill response is sound contingency planning in advance. Existing plans should be strengthened and kept under regular review. In this area, the recently established Marine Emergency Mutual Aid Centre in Bahrain has a critical role to play. It is vital that the Kuwait Action Plan countries see to it that it remains adequately funded to deal with marine pollution emergencies.

Qatar

Qatar is at a crossroads in its national development. Many of the environmental problems and opportunities confronting this small country are shared by sister countries in the region. By building on the positive action taken to date, Qatar can establish an example of sound environmental management which its neighbours will feel inspired to follow.

The export of oil- and petroleum-based products will form the basis of the Qatar economy for the foreseeable future. Industries planned for the Umm Said area and elsewhere present new opportunities for industrial diversification, but also new pollution threats. The overall challenge is to take action now to prevent serious water and air pollution later on. The new industries will introduce a range of unfamiliar pollutants into the local environment, such as phenol, cadmium and zinc. Gaseous outpourings, such as oxides of sulphur and nitrogen, will also increase. The opportunity is to anticipate and build in pollution control technology into the plants now being constructed.

On control of sewage discharges and other aspects of environmental management, as I stated earlier, Qatar has demonstrated its determination to protect the environment. A similar measure of determination is needed to control existing pollution threats, such as the ammonia discharges from the fertilizer plant, sludge dumping on Halul Island, open-pit burning of onshore terminal sludges and ballast discharges.

A fundamental need is to establish a national coordination mechanism for sound environmental management, a mechanism that will provide scope for participation by government bodies, intergovernmental organizations, professional bodies and educational institutions. Qatar could look to a growing community of nations which have established such a mechanism for a blueprint.

Within the overall framework of a new regime in Qatar for sound environmental management, I would like to put particular emphasis on two aspects:

1. first, the role of public information;
2. second, the role of Qatar University.

It is our experience that, unless the public is kept informed of the need for environmental protection, little that is positive will be achieved. The public concern expressed over the 1981 oil pollution incident and now the Nowruz field spillage, not only in this region but abroad, testifies to a profound, and growing concern among people for the quality of their environment. It is their right to expect decision takers to anticipate such problems and to take the appropriate action. In recognition of this need to keep the public fully informed, the Kuwait Plan includes prominent provisions for increasing public awareness.

A top priority for Qatar and its neighbours is to establish public information programmes on all aspects of the environment. The region possesses a sophisticated and wide-ranging mass media network that should be used to maximum effect. With a large expatriate workforce forming a significant section of the population, special attention should be paid to the need to use different languages.

The groundwork has already been laid for a period of extensive cooperation between UNEP and the University of Qatar. We believe the university has the potential to become a centre of excellence for environmental research and studies in the Gulf region. In discussions with the vice-chancellor and his academic staff we have reached an agreement in principle:

● on the participation of the university in UNEP activities at the regional and national levels;
● on the need for greater cooperation with the Environment Protection Committee;

- on the need to establish a postgraduate institute for environmental studies. The institute would offer post-graduate training and a UNEP/UNESCO-supported diploma course for decision makers. The institute would also undertake long-term research.

I am pleased to note that the university will be hosting an important regional training workshop next month on oceanographic sampling, analysis, data handling and maintenance of equipment.

UNEP philosophy

Though UNEP applies a common strategy in drawing up its regional action plans, it is not a straightjacket. The approach recognizes that human civilization is a patchwork of different cultures and economic structures. It recognizes that people have differing demands on the world around them, and differing aspirations, as well as widely contrasting physical environments. The UNEP philosophy is that each society has to learn to manage its own ecosystem. Perhaps this is one reason the Regional Seas Programme has been so successful. Each action plan reflects a region's particular priorities, needs and ability to cope with its special environmental problems.

Certainly, there are hardly any issues, aside from the protection of their shared environment, for which countries like Greece and Turkey, Israel and Libya, Iran and Iraq, and the USA and Cuba would put aside their political enmities and sit down around a table together to reach common solutions to problems, and not just once, but regularly.

In UNEP, we have been immensely encouraged by governmental support for the Regional Seas Programme. But coastal management is only one part of a wide environmental spectrum. Slowing down, and then reversing, the pace of environmental destruction will only come about when governments display a similar resolve and determination to apply the prescriptions we already have to hand.

15

Sustainable development in a developing economy

Address to the International Institute

Lagos, Nigeria, May 1984

Our understanding of the relationship between environment and development has undergone a profound change over the past 10–15 years. At the end of the 1960s it was the conviction of almost everybody that you either had one or the other, that is if you wanted to have development then the price to pay would be a loss in environmental quality.

That view has been overturned as we have come to realize that the two – environment and development – are interdependent. Without environmental protection you cannot have proper development, and without development you cannot have an improved quality of life. International approval for this principle was given in 1972 by the 113 nations attending the UN Conference on the Human Environment in Stockholm. Since then, terms such as 'ecodevelopment', 'sustainable development', and 'environmentally sound development' have come into widespread use. But how much are these terms really understood? How great is the understanding of the complex relationship between environment and economic development? The unfortunate fact is that 12 years after Stockholm, the mass of the general public and a sizeable number of those who make development decisions are not yet fully aware of what is to those in the environment movement an established maxim: long-term development can only be achieved through sound environmental management.

Environmentalists have a duty to prove, and keep on proving, the advantages of environmentally sound development. UNEP's consistent plea since its establishment in 1973 has been 'development without destruction'. And we have justified this by the fact that a strategy of sustained, rational use of natural resources is cost-effective and makes economic sense.

Natural resources are of two kinds: renewable or non-renewable. Renewable resources are those that grow, like fish and forests, or which are replaced by the undisturbed workings of the natural world, such as soil or oxygen. As long as the sun continues to shine, that is to say, to provide us with energy (current calculations put this at around 4000 million years), these renewable resources need never be used up.

Non-renewable resources are oil, coal, mineral ores and so on. These are being exploited at a frightening rate, not through proper use but rather through imprudent practices and irrational use. Used in a productive and non-wasteful manner, most non-renewable resources could last for centuries to come. There is no real hard and fast limit between renewable and non-renewable resources. Renewable resources, if misused, or used at a rate faster than that of their renewal, are also depleted.

'Sustainable development'

The problems posed today by a dwindling supply of natural resources are serious. By the turn of the century, now just under 16 years away, the problems we face today will seem insignificant by comparison. We believe that a world-wide application of environmentally sound development principles will help solve many of these problems.

But we have to say that the term 'sustainable development' has not become the common currency of decision makers. We in the environment movement must shoulder some of the blame. It has become an article of faith, a shibboleth: often used, but little explained. Does it amount to a strategy? Does it apply only to renewable resources? What does the term actually mean? In broad terms the concept of sustainable development encompasses:

1. help for the very poor because they are left with no option other than to destroy their environment;
2. the idea of self-reliant development, within natural resource constraints;
3. the idea of cost-effective development using different economic criteria to the traditional approach (a subject I shall discuss in some detail shortly); that is to say development should not degrade environmental quality, nor should it reduce productivity in the long run;
4. the great issues of health control, appropriate technologies, food self-reliance, clean water and shelter for all;
5. the notion that people-centred initiatives are needed; human beings, in other words, are the resources in the concept.

A consensus on the means to achieve sustainable development has also emerged and can be summarized as follows:

- raising indigenous environmental or natural resource management capability;
- building upon existing experience in the North so as to ensure that lessons are learned from their mistakes of the past;
- ensuring that environmental considerations are not left out in development planning;
- gathering sufficient hard data of an environmental kind (taxonomic, ecological, geological) so as to enable sound development planning to be implemented;
- informing the public of what is at stake;
- concentrating on systems particularly at risk, be it arid lands, watersheds, moist forest or areas of rapid urban expansion.

The need for sound environmental management

UNEP and its partners in the environmental movement are attempting to persuade nations of the short- and long-term benefits to be gained from the sound environmental management of natural resources. Our aim is to see the concepts put to work – put into meaningful action.

We firmly believe that in this way we have a world to win. Much that has already been lost can be regained. We can stop the advancing deserts and re-green at least some of the areas already lost; we can replant the trees and use the forests more wisely; we can combat marine pollution and restock depleted fisheries; we can stop soil erosion and make better use of the land already being farmed; we can prevent air and water pollution and improve the quality of life in mushrooming cities. Above all, we can tackle poverty, a root cause of most of our environmental problems.

The politicians and other decision makers we have to convince are not for the most part people with a background in the physical or natural sciences. They tend to be economists, lawyers, social scientists, bankers, businessmen, planners and the like. Because of their immediate responsibilities, these people do not make decisions with the interests of the next generation in mind. Their decisions are made with a view to the next election, the annual balance of payments or to the next meeting of shareholders. We have to convince the decision makers that an investment in enviornmental protection pays – now. That is the message we have to get through. UNEP has arguments, hard facts and figures to help promote this message. Before I outline some of them, let me briefly touch on some of the concepts which have emerged during the last decade in our understanding of the environment.

Understanding the environment

There is increasing evidence that sustainable development is not realizable unless three critical relationships are recognized and emphasized at very early stages of development planning. The first is the critical relationship between the natural world, which includes human society, and its development. The tangible components of the natural world, forests, atmosphere, soil, the sea and so on, together make up the system which supports all life on earth. It is the strange habit of mankind, one which I have never really understood, consistently to abuse this life-support system on which its survival and future prosperity depend. A series of world conferences on food, population, human settlements, water, desertification and environmental education held after the 1972 Stockholm Conference on Human Environment have served to clarify our understanding of the effects of development activities on the environment, including the availability of natural resources and energy. As a result, a new environmental imperative has emerged: development and conservation of natural resources must be pursued in tandem as goals of equal importance.

The second critical relationship is the connection between development and economic growth. Recent years have witnessed a broadening of our understanding of this relationship. Development is no longer seen exclusively as a matter of the growth rate of national income or the rate of capital formation. The new emphasis is on wider and more qualitative aspects of development, such as improvements in income distribution, employment, health, housing, education, and so forth. One fact that emerged clearly is that an increase in economic growth does not necessarily benefit all sections of society. The conventional wisdom was that an increase in national income would somehow filter or 'trickle down' to the underprivileged. For the most part this has not happened. This is partly because the growth process itself does not necessarily bring about changes in the structure of society. By abandoning the narrow, 'sectoral' approaches to development and promoting instead comprehensive approaches, nations will be able to bring about improvements in income disparities, unemployment, housing, nutrition and health standards.

Third is the critical relationship between our understanding of the interaction between man and nature on the one hand, and the design of practical development policies and objectives on the other. Interactions between forces of change, namely resources, environment, people and development, make it necessary for governments to think in terms of trade-offs between alternative courses of action. Each possible development objective must be carefully evaluated in order to identify its environ-mental consequences. It is important to avoid harmful consequences inflicted by an action in one sector upon other sectors. For example, say an energy department builds a hydroelectric dam, but the forestry department does not follow a sound watershed conservation policy. The result will be a silting up of the dam and a failure to meet energy production targets. A 1979 report by the US Agency for International Development (USAID) observes that deforestation in northern Luzon in the Philippines has silted up the reservoir of the Ambuklao Dam so fast that its useful life has been reduced from 60 to 32 years. This is only one example; there are many others.

Once these three relationships are understood and accepted, environmental principles can be applied to development. And there are compelling reasons to do so with urgency. There is mounting evidence that excessive demands are being made on limited resources and on the carrying capacity of fragile ecosystems; overexploitation of natural resources cannot be sustained. The industrialized nations rely to a large extent on the developing countries as a source of their raw materials, but already the renewable resource base in many developing countries has been seriously undermined. Deforestation, desertification and soil erosion are only a sample of the outward signs that something is badly wrong with the way we approach development.

Integrated approach to environmental protection

The recognition of these critical relationships and the economic hard times of the 1980s both call for a new, more integrated approach to environmental protection. There are two aspects of such an approach.

Improved policy planning

The first relates to improved policy planning. A wide range of new social objectives needs to be identified and defined. Failure to take account of popular feeling in areas which affect public welfare might weaken political systems. On the other hand, greater demands and broader participation must necessarily make the planning process more complex and cumbersome. In this situation new ways have to be sought to meet wide-ranging demands. Narrow sectoral planning will not do. We definitely need a comprehensive approach. This task cannot be accomplished by one single group or department. People in government, in industry, and in the academic field must work together to promote real and equitable social growth.

Assessing the environmental costs and benefits of development

The second aspect of an integrated approach is the urgent need for a realistic assessment of the costs and benefits of environmentally sound development plans. To this end, UNEP developed and is still developing a programme of environmental cost–benefit analysis. Until recently, land and natural resources were exploited without restraint and wastes were discharged freely into the air and water which nobody owned. Because we took the process of resource regeneration for granted, natural resources were considered inexhaustible. However, we now know that the process of self-regeneration can be slow and complex. We now know that air and water have limited assimilative and carrying capacities, so that pollution control measures must be instituted to safeguard the environment and the quality of human life.

It is therefore important, to evaluate the environmental costs and benefits of any development process. But such an evaluation is not easy and is rarely complete. Some of the environmental effects of development can be readily identified and evaluated quantitatively, others cannot. Nevertheless, an economic analysis of the environmental effects of alternative development choices is important because it creates an awareness of the fact that natural resources ought not to be treated as free goods.

Businessmen tend to overestimate the costs of environmental protection, and environmentalists to overestimate the benefits. We need to see the situation in its proper perspective. Such a perspective indicates that environmental protection yields significant economic and social benefits. This is not only in respect of reduced mortality rates and improvements in working conditions, but also in such specific areas as productivity, return on investment, technical innovations, energy use and amenities.

Damage costs

The best available data on damage costs that can be put into monetary terms relate to oil spills in the sea, industrial catastrophes and floods. For example, clean-up costs of oil spills have been estimated at US$1000 per barrel of oil spilled. The total cost of the damage incurred from a huge oil spill which occurred in Japan in 1974, and from its cleaning operations, was about US$160 million. The accident at the Seveso chemical plant in Italy caused damage estimated at US$150 million. The costs of rehabilitation of the American Three Mile Island nuclear power station damaged after the accident in 1979 have been estimated at over US$1.5 billion.

Even without accidents, large quantities of pollutants enter the environment as a result of human activities. Several studies attempt to estimate the economic costs of damage caused by such pollution, including medical expenses, production losses due to illness or death, and agricultural and livestock losses. The cost of air pollution damages in the USA has variously been estimated to be between US$2 billion and US$35 billion per year. Generally speaking, the economic cost of pollution damage in developed countries varies between 3% and 5% of the GNP; this damage cost grew in absolute value from 1970 to 1980, although recently we have observed some signs that these costs may be levelling out. The results of a French study on 24 pollutants indicate that the cost of pollution damage in 1978 was between 3.4% and 4.2% of the GNP. One quarter of this damage was due to air pollution and another quarter to noise. Comparable figures have also been reported in Canada, Italy, and the UK. A recent study in the USSR estimates the cost of health expenditure and decreased work efficiency due to air pollution as the equivalent of US$38 per capita, and the cost of damage to pasture and crops as the equivalent of US$130–135 per hectare.

Damage costs may also be imposed in the process of development through the destruction of certain types of renewable resources. These include the large-scale loss of tropical forests, the degradation of soil due to salinization, the imperfect cultivation of submarginal lands, and so on. More than 30 million square kilometres of cultivated land, with a population of about 130 million, are currently threatened with desertification and consequently with huge economic and human losses.

I should note that this assessment of the costs of damage resulting from pollution and irrational use of natural resources is far from complete. Environmental damage is often selective and unequally distributed in both time and space and among societies. Many of the physical, biological and socioeconomic consequences of large development projects are inadequately known and some cannot be quantified, such as the cultural costs.

Even more difficult is the problem of placing a value on a human life. The traditional economic approach has been to equate the value of a life with the value of a person's expected future earnings. But this approach, apart from ignoring the personal effects of a death which are not measurable in financial terms, undervalues those in society who are underpaid, and it places no value at all on people who are not in income-earning positions.

The costs I have been discussing are damage costs, the costs of not protecting the environment. As I said, they have been estimated to amount to 3–5% of the GNP in developed countries.

Costs of protecting the environment

In stark contrast, the costs of protecting the environment have been estimated to be less than half this amount, or only 1–2% of the GNP in developed countries. Most of this expenditure is for pollution abatement and natural resources protection. In the developing countries, however, the cost of environmental protection is much lower since expenditures are largely limited to drinking water supply and sanitation. Expenditures for pollution control vary from one developing country to another, from 0.5% to 1% of the GNP.

Benefits of environmental protection

The benefits of environmental protection, on the other hand, are highly significant. In the majority of OECD nations, most notably in Japan, Norway and FR Germany, the level of GNP is higher and growth is marginally faster with existing environmental protection programmes than would be the case without them. What emerges clearly is that capital expenditures in the environmental sector make a useful contribution to GNP growth rates.

Expenditure on environmental protection also creates new jobs. A study carried out by the US National Academy of Sciences estimated that nearly 7 million people were employed directly in pollution abatement activities in that country by the mid-1970s. In France, another study indicated that, as a direct result of environmental policies, 370 000 new jobs were created in 1979 alone. Today, in the USA, France, and FR Germany, 1% of the total workforce is employed in environmental jobs.

Additional benefits can be realized when governments coordinate energy and raw materials policies with environmental policy, in order to secure overall savings through the recovery of wastes. Pollution is largely a function of reducing waste, and some private companies are finding that they can make profits by retrieving pollutants. In Switzerland, the USA, the UK, and FR Germany a number of industries have increased their profits, in some instances by as much as 40%, by recycling wastes which would otherwise have been released as pollutants into the environment.

On balance, the costs of environmental policies are usually more than compensated for by the benefits accrued from reducing the damage. The US Council on Environmental Quality estimated in 1980 that the health benefits from a 60% reduction in air pollution in that country would amount to a total annual saving of

US\$40 billion. The US Environmental Protection Agency has estimated that the 12% decrease in particulates alone, achieved between 1970 and 1977, provides US\$8 billion in health benefits each year, compared to the total 1977 expenditures on controlling all air pollutants of US\$6.7 billion.

In the developing countries, the costs of improving the quality of the environment and protecting natural resources are far outweighed by the benefits accruing to society. For example, the construction of drinking water or sewage systems in developing countries could reduce the incidence of infectious diseases, such as typhoid, dysentery, cholera and schistosomiasis by 50–60% or even more. Such an improvement in human health would lead not only to an increase in productivity and time on the job, both of which contribute to an increased GNP, but also to a reduction in expenditures by the medical sector on services and goods, most of which are imported.

Environmental regulations and the ban on production and marketing of some products have had an impact on trade and on the location of some industries. Thus, the relocation of industries to developing countries although mainly based on economic grounds, including the availability of cheap labour and natural resources, can also be based on environmental factors. For example, there is a trend in Japan to locate new aluminium smelters abroad as a means, in part, of avoiding strict domestic environmental regulations. Difficulties in finding environmentally sound refinery sites have forced the petroleum industry to look abroad as well, particularly in Indonesia. In the USA, a trend is emerging towards relocation of industries producing asbestos, mercury, pesticides and other environmentally hazardous substances; for example, asbestos mills have been installed in Mexico and Brazil.

As a result of these trends there will be initial capital flows to the host countries, followed by the shipment of capital goods, accompanied probably by the transfer of technology and managerial skill. There will be positive employment effects. Processing will be carried out closer to the source of raw materials in developing countries, resulting in a rise in the average value of exports. However, such benefits to the economy of the host country could be outweighed by overexploitation of natural resources, together with negative environmental damage and adverse social impacts. It is therefore important that careful assessment should be made of all redeployment opportunities, to determine industrial locations and environmental protection options that would afford advantageous growth conditions and the least degree of environmental degradation.

In sum, the available evidence shows that in most cases improvements in environmental quality have generated significant benefits without negative effects on the economy. The benefits of environmental policies have included reduced mortality and morbidity, improved productivity of labour, technological innovation and increased amenities. But each country, each situation, and each development decision is distinct, so that a continuous assessment of environmental costs and benefits is crucial to long-term sustainable development.

Natural resources and environment in Africa

Mazingira in Swahili, *beeá* in Arabic, and *alá-omá* in Ibo are just a few African words which broadly equate with the English term 'environment'. Collectively, the words mean man's total surroundings and his relationship to those surroundings. These African words also point to an indigenous and historic awareness of the environment.

Age-old customs, practices and religious rituals testify to a concern for the environment among the peoples of this continent long before the 'environmental movement' was given expression and form in the late 1960s and early 1970s by the setting up of organizations such as the UN Environment Programme.

Now, in 1984, we share a perception of what the major threats are to the environment and natural-resource base in Africa. These threats come from opposite ends of the economic spectrum, from underdevelopment and from mishandled wealth. With 20 out of the world's total of 31 least developed nations, Africa has a tiny 2.7% share of the overall per capita income. Underdevelopment is, indeed, the most important cause of environmental destruction in Africa. This continent, through mismanagement, which is frequently manifested in no management at all, is squandering most of its natural resources. But as a major source of the world's raw materials, the nations of this continent are too often the proxy victims of the wasteful development patterns and lifestyles of a number of rich countries.

The conventional analysis is that the major causes of Africa's economic stagnation are the oil price hikes, the spiralling cost of imported manufactured items relative to the prices paid for raw materials and cash crops, and the deflationary conditions attached to development assistance loans. Of course these factors are all very important, but it is my contention that they have received too much attention, while the environmental dimension to economic impoverishment has been overlooked.

Destruction of the environment

If Africa is the only continent where food production is being outpaced by population growth, if environmental diseases are still claiming millions of lives, if deserts are spreading as fast as ever, and if the quality of life in African cities and villages is getting worse, not better, it is because the age-old respect for the environment has been abandoned in the pursuit of inappropriate patterns of development and lifestyles.

In some West African countries the farmers who once made those countries self-sufficient in food have turned to cash crops such as peanuts and cocoa. In East Africa, nomad pastoralists have been squeezed into a marginal rump of their former rangeland, and in Southern Africa, apartheid and the Bantustan policies have moved indigenous peoples to areas with the least agricultural potential. In all these cases, the small farmers and pastoralists have been shifted to marginal lands which are unsuited to their traditional cultivation practices. The result is environmental devastation in the form of soil loss and desertification. It may have taken nature a thousand years to accumulate soil on these lands to a hoe's depth. Overgrazing, overdeep ploughing, and needless removal of the indigenous protective vegetative cover have in some cases razed that precious mantle of life-giving soil in a manner of months.

If African nations permit these kinds of destructive processes to continue, they will assuredly look back in 10–15 years from now and envy the economic problems of today. In looking back, the next generation of decision makers will want to know why we did so little when we were aware of the scale of the problem and when, more importantly, we had the solutions in hand.

The need for environmentally sound development

I would like to indicate why, and how, the African nations should begin in earnest a policy of environmentally sound development. Africa is confronted by an acute firewood shortage. About two-thirds of Africa's energy south of the Sahara is met by firewood. Already a country like Somalia has to all intents and purposes run out of firewood reserves, Kenya will do so within the next 10 years at present rates of exploitation, and even countriess in the moist, tropical belt, such as Ivory Coast, can expect to lose their remaining tropical forest by the turn of the century. That is, if we carry on as we are.

Notwithstanding the expectations we have for renewable energy development, Africa must come to terms with the fact that for the next 20–50 years trees will be the major source of energy for the poor. Recognizing this, the 1981 UN Renewable Energy Conference in Nairobi called for a five-fold increase in tree planting. But to reach such targets we need to involve people in imaginative schemes and the authorities have to help them make the right decisions about which tress to plant: fast-growing trees, yes, but trees that at the same time are nitrogen fixers, can withstand droughts and can provide animal fodder and food for people.

Africans are rightly angry when they are told by developed nations that they are overpopulated. Nevertheless, with predominantly agricultural economies often based on relatively poor soils, and with an overall continental population increase currently reckoned to be nearly 3% each year, family planning is an indispensable part of achieving sustainable development. At current rates of increase Nigeria, according to one estimate, will have a population midway through the next century of over 600 million.

Fewer mouths to feed each year will enable governments to focus on the problem of improving the quality of life of those already born. This involves addressing one of the most serious problems facing African countries, namely the provision of water, not only in more plentiful supply, but also water that is clean. In the rural areas of the developing world, more than 70% of the people must drink and wash with dirty water.

A sound environmental policy is inseparable from the campaign to prevent the comeback of waterborne environmental diseases. In tropical Africa, for example, 50% of children up to the age of three are affected by malaria, which kills one million people every year. The construction on a continent-wide scale of hydroelectric power and irrigation schemes that ignore environmental considerations has been an important factor in accounting for the persistence of bilharzia, which affects around 200 million people.

I believe that some of Africa's environmental problems can be traced to a persistence of the old colonial idea that it is still an empty continent. Against this background, independent nations have been too eager to give the go-ahead to major development projects like million-acre cotton farms without giving due attention to potentially serious environmental side effects, such as waterlogging, waterborne diseases, salinization, destruction of vegetative cover, water pollution and soil erosion. With built-in environmental considerations, such grand schemes would certainly be profitable, although smaller ones utilizing improved existing cultivation methods might be even better money makers.

More attention should be given to making better use of land already being cultivated. We know that for every hectare of land coming under new irrigation in the Sahel, one hectare of already irrigated land is going out of production due to mismanagement.

Simultaneously, African countries should pay much more attention to implementing already devised methods to prevent post-harvest losses. In Africa, almost 30% of all crops – enough, it is estimated, to feed more than 50 million people – are lost in storage every year.

Drought

A complicating factor in achieving sustainable development in Africa is drought, which has affected the continent for a long time and is getting worse. We are witnessing a drought which is more widespread than that of the late 1960s and early 1970s; a drought which may not yet have run its course. We are seeing around us, in various parts of Africa, disaster and famine. Scenes of famines in Ethiopia, Botswana, Morocco and Mozambique and of dried-up dams in West Africa once again graphically show the vulnerability of people living in the world's drylands. It is true that the Secretary General of the United Nations is mobilizing the whole UN System, UNEP included, to face up to the crisis. He is calling on donors to meet the emergency. But we are all aware of the fact that the solution lies in dealing with the causes of the crisis. One essential element is to understand and accept the fact that drought is recurrent and cyclic. If it goes away now, it will come back. It is therefore crucial to learn how to live with the drought, and how to plan for development with drought always high on our minds.

The need to preserve natural resources

It is a mystery to me that, while governments consider oil, industrial plants, mines and so on as part of the national assets, soil, clean air, clean water and intact watersheds are not included in any accounting system. Unless and until an environmental audit system is systematically applied we will continue to abuse our resources, a process which was recently described as 'biological deficit financing'.

The destruction of the world's genetic resources – our wild plant and animal species – is an outstanding example of biological deficit financing. Scientists need genetic variety for cross-breeding to produce higher-yield crops and to keep one step ahead of plant and animal diseases.

In my own country, Egypt, a great library was created in Alexandria many centuries ago. It was a repository of the accumulated knowledge and wisdom of the Mediterranean civilizations. It was burnt to the ground and with it went that great store of writings, lost for good. In a sense that is what we are doing with the genetic pool, but it is a smouldering fire which we are doing very little to extinguish: 10% of all flowering plant species are threatened with extinction, and of all the 'books' in our plant library, just 1% of the higher plant species has been studied.

Preserving genetic resources, as with everything I have spoken about today, comes down to good environmental management. At the local level it means regulating firewood collection and grazing in a buffer zone around a protected area; at the national level it requires setting up and policing a network of protected areas; at the international level it is for organizations like UNEP and FAO to do all in their power to make sure the developing nations, which possess 65% of all species, get a fair return in the international seed market.

Call for action

In conclusion, I bring you a message of warning, a message of hope and a call for action. The warning is that environmental deterioration, through its impact on the economy, is also undermining peace and security. The Brandt Report put it clearly: 'Few threats to global peace and survival of the human community are greater than those posed by the prospects of a cumulative and irreversible degradation of the biosphere on which all human life depends'.

My message of hope is that a number of governments are making preparations for a quantum leap to stop and reverse environmental deterioration. Several African countries are leading the way by drawing up national conservation strategies, passing regulations requiring all new development projects to be screened for environmental soundness, starting tree-planting campaigns and so on. Collectively, African nations have endorsed environmentally sound development in the Organization of African Unity's Plan of Action agreed here in Lagos.

And now my call for action. For thousands of years, African villagers and townspeople have used local resources prudently so that they have been handed over intact from one generation to the next, often unconsciously as these people observed religious, cultural and other practices finely honed to their environment. In terms of material possessions, these people were poor, but in their respect for what nowadays we refer to as the balance between people, resources, environment and development, they were rich indeed.

What I am calling for is a restoration of that balance. This does not mean I am calling for a brake on modern development through a return to a pre-industrial economy. Rather we must apply alternative and less wasteful development patterns which will revive and, where it has already disappeared without trace, resurrect a respect for living resources. Simultaneously, governments must do all they can to bring the material benefits of the most advanced scientific and technological developments to increase the standard of living for all their people. In short, we must combine the best of the old with the best of the new.

16

Desertification is stoppable

Statement to the Twelfth Session of the Governing Council of UNEP

Nairobi, Kenya, May 1984

Seven years ago in Nairobi the UN Conference on Desertification, attended by 94 nations and 73 observer groups, agreed a Plan of Action to Combat Desertification. By general agreement, the Conference is considered to have been one of the best prepared UN conferences of the 1970s from a scientific standpoint. The Plan's 28 recommendations were ambitious in scope but entirely in keeping with the extent and complexity of the problem revealed by the studies prepared for the Conference. Effectively applied, those recommendations would have put the international community in a position to have halted desertification by the year 2000.

Later on in this meeting the Governing Council is to review the current status of desertification and the progress made in the implementation of the Plan of Action to Combat Desertification. UNEP's Desertification Branch has coordinated a comprehensive assessment of the period from 1978 to 1984. The main findings and priorities for action are contained in my report to this 12th Session of the Governing Council.

We admit that our assessment of the status and trends of desertification is based on data which, for a number of regions, are not entirely adequate. When in doubt, the experts who conducted the survey erred on the side of caution. However, the facts are clear. Those who held such high hopes after the Desertification Conference that we would now be on course for achieving the 1977 goal of stopping desertification by the turn of the century will find the report depressing reading. The assessment finds no signs that the war against the spread of desert-like conditions is being won. The goal set by the Conference of stopping desertificaton by the year 2000 must now, therefore, be considered an unrealistic target.

Drought and famine

In 1977, we met with the appalling scenes of starvation and destruction in the Sahel fresh in our minds. This year we meet against the background of a new, more widespread drought which may not yet have run its course. Scenes of famine in Africa – dried up dams, roads and railways inundated by sand, and slaughtered livestock – have once again graphically shown the vulnerability of people living in the world's drylands. In Ethiopia three million people are experiencing famine, in Mozambique more than two million. World-wide, nearly 60% of the rural inhabitants of the drylands are threatened by desertification.

As the Desertification Conference recognized, drought only exacerbates the insidious process of desertification. There can be no doubt that, if national governments and the international community had applied the Plan's prescriptions, the human suffering caused by the last year's rainfall failure would have been far less.

The message now is the same as it was in 1977: when we cope with desertification we can cope with drought. The difference between now and then is that now we have no excuse. The framework for applying the known remedies for desertification has existed for over six years; the suffering of the millions of poor in the semi-arid regions should be on all our consciences.

Effects of desertification

In the seven years since the Conference we have learned that desertification leads not only to a decline in productivity, but also to an increase in atmospheric dust, destruction of natural water recycling and drainage systems, loss of the genetic diversity so necessary for maintaining industrial and agricultural productivity and loss of markets. Ultimately desertification pushes villagers off the land and into the already severely overcrowded slums and shanty towns, or onto yet more marginal lands. Still others are forced over national borders to seek a better life. We have begun to call these people environmental refugees.

Especially alarming are signs that desertification is feeding itself. Loss of tree and other vegetation cover could lead to a drying out of local climates by increasing heat reflectivity and reducing the moisture content. The sum of local climatic change may be one reason why the drought in the Sahel which began in 1968 still persists. Last year was the worst on record.

Currently, land irretrievably lost to desert continues at 6 million hectares per year, and land rendered economically unproductive is showing an increase over the 1980 estimate at 21 million hectares per year, an area approximately the size of Bangladesh or more than one third of the size of Kenya. We estimate the cost of not stopping desertification, in terms of loss in agricultural production over a 20-year period, to be in the region of US$520 billion.

We can, of course, put no price tag on the toil desertification is taking on the 850 million people whose livelihoods are now considered at direct risk. The rural population in areas now known to be severely affected has increased from 57 million in 1977 to 135 million in 1984. It is true that, in part, this jump in the figures arises from the addition of sub-humid tropical areas now known to be affected by desertification, but there is also a definite increase in the number of people affected in the dryland areas included within the 1977 definition.

The despair and hardship which desertification engenders cannot be quantified. Merely by trying to feed their families, the poor who live in the world's drylands and sub-humid tropics are locked into a cycle of destruction. They are bewildered and angry and they are looking to governments, the UN and other organizations represented at this Council for the solutions.

Inadequate response to the Plan for Action

Our assessment confirms that the Plan of Action itself remains viable. Despite UNEP's continued warnings of the threat posed by desertification, the response by developed and developing countries alike has been totally inadequate. Many of the Plan's recommendations remain a dead letter. Virtually nothing, for example, has been done to implement the transnational projects and, as of today, the Special Account set up by the General Assembly in 1980 has received only US$150 000.

The failure to apply the Plan of Action in any meaningful way is only one casualty of a clear retreat from the degree of multilateral cooperation which the UN helped to inspire. In his most recent report on the work of the organization, the Secretary General said that this backslide must be reversed if 'we are to avoid chaos and disaster on a scale hitherto unknown'.

Your consideration of the findings of the assessment of desertification, however, should not be distracted by recrimination, or by futile debate on who is responsible for what. After all, all of us are responsible. The opportunity is to learn from our mistakes, to identify the causes and to inspire a comprehensive and effective commitment to defeating desertification. A major responsibility therefore devolves upon this Council. We must leave governments and the general public in no doubt that we are confronted by a process whose destructive effects are comparable with a nuclear weapons exchange; slower perhaps, but no less damaging to people and the web of life that supports them.

Adoption of a Programme of Action

A headline recently appeared in a major western newspaper which said 'March of Deserts Unstoppable'. This fatalistic message is not the one we wish to convey. Our assessment has turned up enough evidence to show that when the plans are realistic they are workable and that, when properly motivated, people can combat desertification. Our basic task, therefore, is to adopt a realistic Programme of Action for the next 15 years that is equal to the increasing dangers that now confront the more than 90 member states of the UN affected by desertification. I have done my best to assist you in this important task through the proposed action at the end of the report before you. We must communicate in the clearest form possible the benefits to be gained from stopping further desertification and rehabilitating land on the verge of turning to barren sand.

Economic self-interest, compassion and concern for the fate of the coming generations are the motivations. The resources that need to be mobilized are not massive; the current estimate of the investment required each year is US$4.5 billion. The total external assistance to those developing countries in need of such assistance amounts to US$2.4 billion. The cost of neglect has been shown to be five times greater than the total cost of action. The bill would amount to a tiny fraction of what we will spend on arms between now and the end of the century: US$70 billion in efforts to halt desertification, and thereby construct more favourable conditions for life; US$15 000 billion for producing and purchasing arms to threaten our life-support system.

Nor need we think only about traditional aid schemes. The governments have had before them for several years now, through the General Assembly, a range of suggestions, including the creation of an Independent Financial Corporation, which could help provide the additional resources so desperately needed. Such arrangements would also guarantee a measure of automaticity and predictability in the provision of funding. The response has been disappointingly slow. All governments which have not yet done so have been requested by the General Assembly to inform the Secretary General of their views on these suggestions to allow the General Assembly to take a definitive position by the end of next year. The Consultative Group for

Desertification Control is a viable mechanism that needs urgent support and strengthening. I am proposing another expansion of its mandate and a better use of its potential.

Mobilizing people to help themselves

Another lesson we have learned is that while the problems are on a dauntingly large scale, the response should not only be at the level of governments, but also at the level of individuals and small communities. Unless the people most affected are helped to help themselves, the battle against desert encroachment cannot be won. Some non-governmental organizations have been particularly successful in mobilizing local communities to take action. Our assessment shows that the effectiveness of small-scale projects focused on local problems compares very favourably with the larger, more costly downward-directed schemes promoted by governments and development agencies. We are proposing that UN agencies should do much more to provide matching funds to get a greater number of such schemes and projects underway.

In the Indian state of Gujarat hundreds of thousands of villagers and town dwellers have been involved in a successful social forestry experiment. In Ethiopia, the UN's Food for Work programme has involved similar numbers in soil terracing. Such successes give us hope and show the way forward.

Causes of desertification

A fourth point comes over clearly from the assessment, providing a signpost to future action. Unlike most environmental problems, no uncertainty surrounds the causes of desertification. Outwardly they lie in the misuse and overuse of fragile lands – the overgrazing, deforestation, overcultivation and poor irrigation that destroy the biological productivity of the land. These practices result from a complex mix of hidden causes, which include unfair terms of trade, the increase in human numbers, short-term planning and, sometimes, just sheer incompetence. The recommendations made by this Council must take these hidden causes into account.

Constraints on tackling desertification

A major constraint identified by the assessment is the low priority developing countries give to fighting desertification. A new willingness on the part of their governments to help the rural poor, who in the remote drylands, tend to be politically feeble, would undoubtedly encourage development assistance agencies to change their priorities too. Though the destabilizing effects of desertification may show up ultimately in the towns, it is in the countryside that the battle will be won or lost. Thus, a top priority is to channel more effort into conserving the 3435 million hectares of rangeland and rainfed cropland already affected by desertification. Greater attention should also be given to countering the adverse consequences of irrigation schemes. The assessment finds that 40 million hectares of irrigated land are affected.

Another major constraint has been, and continues to be, the lack of resources. The failure to make funds available lies in the reluctance to plan for the long term.

Investment in antidesertification, by its very nature, can only produce comparatively modest, though largely secure returns over a period of decades. Overcoming this constraint depends on a new preparedness by governments to change their time horizons for planning.

With so many apparently more immediate and pressing problems, governments may feel they cannot afford to concentrate on long-term development projects like range management and agro-forestry which produce so few exportable surpluses. If this meeting is to be a success, it must show why they cannot afford not to.

National plans of action

The seventh element of the concerted approach that we need to adopt is the urgent need for governments to draw up national plans of action, as recommended by the Desertification Conference. Only three governments have done so. They must also set up the national machineries to coordinate action. Only one has done so.

National plans to combat land deterioration should be merged with plans for economic and social development, but at the same time should have a clearly defined place and allotted resources. A properly funded and strategically placed national machinery would act as a forcing house in the overall governmental process, ensuring that all development plans take account of desertification. In would also act as a point of leverage for raising additional funds from donor nations and multilateral organizations.

Regional cooperation

The last and probably one of the most important ingredients for success is more regional cooperation. Like most environmental problems, desertification pays no heed to national boundaries. My report recommends, for example, the introduction of a mechanism along the lines of the UN Sudano Sahelian Office for the nine southern African countries which have been so hard-hit by the recent drought.

Desertification is included in this year's State of the Environment Report as an area ripe for cooperation between and amongst developed and developing nations. India's success in coping with desert creep in Rajasthan, for example, could have a wider application in other developing nations. So too could PR China's success with irrigation. What I am hoping is that the developed nations which are experiencing their own desertification problems will be especially sympathetic to the plight of developing nations whose capacity to tackle the problem is so very much more limited. A new era of cooperation on the environment could help give new purpose and direction to the dialogue and cooperation between North and South.

Perceiving the gravity of the problem

In the final analysis, the preparedness of the international community to tackle the global threat posed by desertification depends on a perception of the gravity of the problem; on an understanding that the political and economic security of all nations is already being undermined by desertification. There are some encouraging signs that this is happening. For example, each head of state attending the recent meeting in Niger of

CILSS, the Permanent Interstate Committee for Drought Control in the Sahel, identified tackling desertification as a top development priority. The CILSS meeting has passed on a special appeal to the June meeting in London of the leaders of the West's major countries. We must hope that they answer the call and channel more assistance through their organization, the Club du Sahel.

I feel that the recommendations and priority areas for action set out in Part 3 of my report provide the framework for revitalizing the Action Plan, as well as areas for useful cooperation between and among developed and developing countries. All represent vital components but I would, in particular, emphasize the need:

1. to improve greatly land-use planning in drylands;
2. to ensure that priority attention is given to land of highly productive potential considered to be at greatest risk;
3. to concentrate on improved water management, an aspect of environmental management which rivals energy in importance;
4. to give more emphasis to training and applied studies to improve national antidesertification capabilities;
5. to go down to the level of small community projects and programmes;
6. to overcome the political constraints to improved regional cooperation;
7. to launch a meaningful effort at the international level.

This Governing Council must set the alarm bells ringing and approve a Programme of Action that will be workable. UNEP has done its best to mount an information campaign that will ensure the public is aware of the issues. For it is only when the tiny minority in government and among the public who now perceive the threats become a majority that the battle against the deserts will be won. Desertification is threatening to go out of control; your opportunity is to spark off a new endeavour to stop that happening.

17

Environmental impact assessment – principles and guidelines

Statement to the Working Group of Experts on Environmental Law:
First Session on Environmental Impact Assessment

Washington, DC, USA, June 1984

This occasion marks an important step in the effort to establish legal principles for incorporating crucial environmental values into the development process. The potential long-term value of this work is hard to overstate.

I am most satisfied to see that the process of developing specific global principles and guidelines on environmental impact assessment (EIA) is beginning in the USA, since this country was the first to put the EIA process formally into effect through the National Environmental Policy Act of 1969. Soon afterwards, at a meeting of 113 nations in Stockholm in 1972, a parallel international resolve emerged when these nations firmly stated their common conviction that rational planning constitutes an essential tool for reconciling any conflicts between the needs of development on the one hand, and the needs of the environment on the other.

This resolve has taken root and grown over the last decade. At the last count, 17 nations had adopted formal EIA procedures; four more nations had adopted EIAs as part of their development planning processes, and a number of others – including the entire European Economic Community – are now in the final stages of developing EIA procedures. But the environmental impact assessment process has not yet been accepted in many parts of the world, and it has yet to become a significant part of development planning in some of those countries where it has been accepted. I am concerned at the slow pace of acceptance, for it is a disturbing indicator that, even now in 1984, developmental values and environmental values are not being reconciled in many parts of the world.

Shifting development process

The rational linkage between development and the environment is a crucial one. Each year the world's population grows by about 80 million people. While the world's population has been increasing, the average per capita income in many developed and developing countries has been rapidly rising, which in turn has created an unprecedented surge in demand for renewable and non-renewable natural resources. Moreover, the disparities of income among countries, and the shifting patterns of these disparities, have resulted in locational shifts in demand for these renewable and non-renewable resources.

Thus, not only do we have a continuous development process, but one which shifts over time. Relocation of industry is a good indicator of these shifting patterns. According to data compiled by the United Nations Commission on Transnational

Corporations in 1978, transnational corporations were investing at that time one quarter of their foreign assets in developing countries, at the rate of about US$290 billion per year. Each dollar invested, each plant built – whether in a developing country or in an industrialized nation – comes with its own particular environmental value, either positive or negative. The EIA process gives decision makers around the world a rational way of determining that environmental value, and of reaching more enlightened development decisions as a result.

Environmental impact assessment process

Occasionally I have seen policy makers in different parts of the world meet the EIA process with fear and scepticism: fear that development and growth might be hindered, perhaps; or scepticism that the process can be practical only in particular applications. These reactions, while understandable, underline the supreme importance of your mission. By establishing clear, understandable and sensible principles of law, you will provide the foundation for a clearer understanding of the decision-making process around the world, and establish a practical working framework for routinely considering environmental values in the development process.

But a word of caution: complex or lengthy environmental impact assessments are of little use to policy makers. Complex EIAs may in fact be detrimental when misused and may hinder, rather than enlighten, the process of development. A number of countries have introduced a 'scoping' step in the EIA process, to delineate the scope of an EIA at the outset, thus focusing limited human and financial resources on the environmental areas of highest priority. Above all, I urge you to remain practical in your deliberations, never losing sight of exactly who makes development decisions, and how the EIA process can be used in practice by these individuals to make better decisions.

I also urge you to make sure that the principles you develop are flexible enough to be adapted appropriately to diverse situations. An appropriate EIA for Indonesia may be very different from an appropriate one for Luxembourg, when socioeconomic sectors are taken into account, although many elements of the process may be similar. To have a positive and long-lasting effect, environmental values must be determined and incorporated into the development process from within a government, not imposed from without.

Principles and guidelines for environmental impact assessment

Each of you has been invited here today as a direct response to the combined wills of nations. The Working Group of Experts on Environmental Law, established in 1977 by 26 member states of UNEP's Governing Council, alerted the community of nations to the need for legal guidance and principles in the field of the environment. This led to an important meeting on environmental law in Montevideo in 1981. One of the specific decisions of that meeting was to develop principles and guidelines for environmental impact assessment, a decision which was subsequently embraced by the member governments of UNEP's Governing Council in 1982 and 1983.

Each of you is therefore a part of this international consensus, in that you are its executors. You are the people who will translate broad mandates into specific legal actions; you are the people who will make the environmental process work on a global scale.

With this in mind, I strongly urge you each to listen to individuals who have had hands-on experience with environmental impact assessments; individuals who have written them, decision makers who have used them, people in industry who have been affected by them. I urge you to listen to individuals from developing and developed countries alike. These are the people who have learned from past mistakes. These are the people who know how to make the EIA work in practice. These are the people we need to support our effort to establish environmental impact assessments as an integral part of the world's development process.

To aid acceptance of the environmental impact assessment process, a simple, practical and cost-effective format for EIAs is needed, which could be used by all countries without fear of undermining institutional or financial resources. There have been several recent attempts at such a simplified approach, which you may wish to consider during your deliberations.

Another point which you may wish to consider is the extent to which the proposed principles and guidelines you develop should focus on the transboundary environmental impacts on development activities. These transboundary impacts are becoming increasingly common and, as I have just witnessed at the international acid precipitation meeting in Munich, such impacts are extremely difficult to rectify after the fact. As a consequence, the environmental impact assessment process is a most appropriate mechanism for considering in advance the possible transboundary environmental effects of an activity. Several international legal instruments already exist which call for the elaboration of state obligations for assessing the transboundary environmental impacts of certain activities. While international treaties are not immediately within the scope of your charter, I note that principles and guidelines are valid expressions of international will, and may well form an appropriate basis for treaty making at a later stage in the process.

The work confronting you at this meeting is demanding, but I believe you will find it rewarding. I assure you of UNEP's total commitment to the cause which has brought us together, and I wish you every success in your deliberations.

18

World population increase – our dangerous opportunity

Address to the Second International Conference on Population

Mexico City, Mexico, August 1984

If there had been only a 0.5% annual increase in the world's population since the time of the Roman Empire, the total number of human beings would be a thousand times greater than it is today. Until the onset of the Industrial Revolution and the advances in health care, high rates of mortality kept population in check. Man's success in conquering many diseases and in providing a greater variety and quantity of food has lowered mortality rates. So when we consider the problem of how to deal with the population increase it should be remembered, first and foremost, that the recent increase in human numbers is a product of our success.

However, that success has been relative. When two out of every five people who die today are children under five, when life expectancy in parts of Africa is scarcely more than the age of 40, we have little cause for complacency. Any policy to slow the birth rate must continue to be accompanied by campaigns to lower mortality.

In 1984 there are alarming signs that the yearly increase in world population – currently some 1.7% – is undermining our very considerable achievements in providing for the needs of all peoples. Each year more than 80 million people – roughly the combined population of the UK, Sweden, Ghana and Ecuador – are added to the total. Every five days the world's population grows by a million people. Already, the resources and natural systems which support the world population are showing signs of strain, severely in some cases. Deforestation, soil erosion, desertification, air and water pollution are only the outward signs that we are mismanaging our resources. Of all those resources, UNEP holds that people are the most valuable, a precious resource that needs to be managed too.

Given the present age structure and the current rate of increase, I feel we must plan for a global population of about 8 billion by 2025. Midway through the next century, when the population is projected to level out at over 10 billion, nine out of every 10 people will live in the developing countries. The population of the Indian subcontinent is projected to stabilize at over 2 billion, and Nigeria's at 600 million – equivalent to the current population of the African continent.

The sheer scale of human numbers will put increasing strains on the environments of countries now experiencing annual growth rates of between 2% and 4%. Against this background the temptation is always, and rightly so, to resort to the term 'crisis'. It is interesting to note there is no direct word for crisis in China – instead it can be translated as 'dangerous opportunity'. In UNEP this how we view the particular role of slowing population increase in the overall endeavour to strike a balance between people, resources, environment and development. Curbing population increase presents great opportunities, failing to do so presents dangers in proportion.

In the period since the first World Population Conference in 1974, increasing importance has been attached to studying the links between population and other areas of global concern. The population dynamic is no longer seen in isolation. Population planning is increasingly seen as one component of the overall process of economic and social development. UNEP expects that this new understanding will be fully reflected in the Plan of Action which will emerge from the second Population Conference.

The about-turn

There are parallels between the 1972 Stockholm Conference on the Human Environment and the Bucharest 1974 World Population Conference: both were convened primarily as a result of pressure from the industrialized countries to tackle, in the case of the former, industrial pollution and, in the latter, overpopulation in its narrow demographic sense.

Developing countries wondered about the validity of the term 'overpopulation' when applied to them when most European countries at that time were more densely populated than most developing countries. Instead, they saw the problem as one of use of resources, where the average person in an industrialized country used 20 times more resources than his neighbour in a poor developing country. The issue is not simply one of numbers but equally one of how resources are utilized. In considering the latter, the question of individual resource utilization is a primary issue, both in developed and in developing nations.

Any plan that seeks to curb population increase is meaningless unless it addresses both the vast differences in resource use and the issue of sustainable resource management. I still believe this central point was best summed up by Mahatma Gandhi who, when asked if he would like an independent India to be like Britain replied 'if it took Britain half the resources of the world to be what it is today, how many worlds would India need?'.

Since the early 1970s there has been a remarkable about-turn in attitudes. The developing countries are now in the forefront of the move to conserve the environment. They see clearly the role of conservation in increasing productivity, and there are very clear signs that they are becoming equally supportive of the need for national population policies. Partly as a result of UNEP's work on inter-relationships, a consensus has emerged that effective environment and population policies are crucial factors in achieving sustainable development.

It is only through wise and rational use of resources that we can achieve the standard of living necessary to strike a balance between people, environment, resources and development. Reaching a demographic balance is at the same time both a means to this goal and a result of achieving it. The publication of major documents like the Brandt Report and the World Conservation Strategy have been important bench-marks in the emergence of this consensus. UNEP is confident that the Plan of Action which will emerge from this conference will be another.

Resources in peril

In 1982, on the eve of our 10 years commemorative conference, UNEP published a major scientific survey which showed that, with very few exceptions, renewable resources are already seriously overstretched in many regions of the developing world. At the global level, few generalizations are meaningful, but at some regional and national levels, the speed and rates, for example, of soil erosion, open and closed tropical forest destruction, exhaustion of freshwater supplies and species extinction are a cause for great alarm. We may not yet have reached crisis point, but the dangers are becoming more obvious and the opportunities fewer. Even taking into account the anticipated advances in science and technology and what can be achieved through human ingenuity, the situation has, and is, becoming perilous over vast areas.

Take desertification which has been exacerbated by the recent world-wide spate of droughts. UNEP has recently completed a two-year global assessment which finds that each year some 6 million hectares are turned into desert. A further 21 million hectares of farmland – an area equivalent to Bangladesh – are reduced to zero economic productivity. Upwards of one third of the world's arable and grazing land, which supports 20% of the world's population, is at risk. The causes and consequences of desertification are complex. Problems of management, land tenure, climatic trends, government policy and rural neglect are all involved. The increase in population is only one element in the equation. The consequences of desertification go beyond loss of productivity to include increase in atmospheric dust, loss of genetic resources, drying out of local climates, disruption of natural drainage systems and loss of markets.

In the Sahel the human population has doubled over the past 25 years and is on course to double again by the year 2000. Given the low level of development in the Sahel, satisfying the Sahelian peoples' basic needs will mean an increase in livestock and crop production which, if improperly carried out, will mean cultivating yet more fragile dryland. Last year's failure of rainfall left the Sahel countries 1.6 million tonnes short of the food grain needed for bare subsistence, and yet one drought-hit Sahelian country, through apparently more effective management, was able in 1983 to feed itself and even to produce a small surplus. World-wide there are enough of these successes to confound the pessimists.

People, resources, environment and development

One important priority for you should be to add a new section to the Plan on the interrelationships between people, resources, environment and development. The existing Plan contains little that deals explicitly with these interrelationships. The nature of the resources available, the demands of the people, their cultural, religious, aesthetic and other needs differ hugely, not merely from one region to another but from one locality to another. The interactions between, say, an Andean peasant, a Sahelian pastoralist or an American wheat farmer and their respective environments are very different.

The methods of resource utilization vary according to circumstance. Resources are the main components of the environment, their utilization is the process of development. Misuse of resources by people leads to deterioration of the environment. In turn, this deterioration represents the non-sustainability of the process of development and hence undermines the meeting of people's needs and aspirations. The interrelationships are complex: each of the four components affects and is affected by the other three. To be relevant to our concerns in each of these areas, our policies and actions must be relevant to this system as a whole.

The need for a system-wide approach

UNEP has sought to develop our understanding of the life-support system as a whole. We encouraged the holding of a UN symposium on interrelationships in Stockholm in 1979. UNEP followed up by convening two meetings of a high level expert group on the subject in Nairobi in 1980 and 1981. In brief, we sought an increased understanding of how the system interacts. This would help in the identification of points of leverage where coordinated action could result in positive, mutually reinforcing and accumulative results.

An essential aspect of achieving positive results was to stress the role of people as effective participants in, and beneficiaries of, the development process, stressing their role not just as consumers of resources but as human resources. It was agreed that the development of a conceptual framework based on interrelationships required a much more comprehensive understanding of the complexities. Further knowledge of interrelationships would come through the implementation of a programme of work, involving case studies, and applying existing knowledge. Such a programme was developed and agreed, with priority being given to two case studies, one on deforestation of the Himalayan foothills, and the other on overgrazing in the Sudano-Sahelian region. An institutional framework was devised and its later acceptance by the General Assembly led to the establishment of a Trust Fund to finance the programme of work.

With UNEP carrying out the coordinating role, the UN system will shortly put into operation the case study on environmental deterioration in the Himalayan foothills. Soil erosion is occurring on a massive scale. Increasing human numbers have increased the demand for firewood and land for cultivation. Other causes include road building, the demands of tourist trekkers, timber concessions and similar deliberate development activities. The network of Himalayan ecosystems is being subjected to mounting strain. Landslides have become common and the hardships of the local people have increased. There are reports of suicides among village women who could no longer endure the additional burdens. The effects of deforestation and soil erosion in the higher lands are felt in the great river systems of the Indus and Gangetic plains below, where silting rates and river flow patterns are important factors to the livelihood of hundreds of millions of people.

The connections between the pressure put on the forests by growing populations, the erosion of the economic basis for the survival of people in the hills, and catastrophic flooding in the plains are becoming clearer. So too is the nature of the measures required to deal with the problem, of which population planning must be a part. The need to get such efforts underway by the countries concerned is now felt strongly by the relevant institutions of the UN system.

Our expectation is that these studies will clear up the divergent and confusing results of analyses based on apparently straightforward two-way combinations. Thus on the one hand we have some analysts reaching the conclusion that in some areas the carrying capacity of the land has already been exceeded. Some of the macroanalyses of carrying capacity conclude that an increase in population at the current rate will prove disastrous. For example, it has been pointed out that in Bangladesh in 1981 each hectare supported more than six people. What chance then for adequately feeding triple that number? At the other extreme, some studies conclude that continued moderate population growth will be beneficial. An FAO survey on potential food production projects a possible global balance between population and food supply well into the next century.

Fixing carrying capacity

These wide differences stem from the difficulties involved in determining what exactly is the land's 'carrying capacity' as a resource. It is not a fixed concept; it changes as discoveries are made, new technologies are developed and applied and techniques of resource management are improved.

The matter is further complicated by the fact that even studying a combination of two elements is not a linear exercise. For example, in considering the population–resource balance, the long-term challenge of feeding growing populations involves a complicated mix between numbers and demand for food on the one hand, and on the other resources, levels of farming technology and ability to pay for food imports by exports of commodities. Each factor must be assessed carefully and an attempt made to bring population growth and possibilities of food provision into a rough balance.

A recent FAO/UNFPA survey found that, with a continued low level of farming inputs, 64 countries with a projected turn of the century combined population of over one billion will be unable to produce enough food to feed their increased populations adequately. Even with an intermediate input level, 36 countries will be in a food deficit position.

In the current state of agricultural technology, raising the level of inputs will also lead to an increase in demand for fossil fuels. Growing and packaging food for the average European and North American uses the energy equivalent of 10 barrels of oil per capita each year. If the present world population were all fed in this way, the world's proven oil reserves would be exhaused in a mere 11 years – a fact which underlines the need to develop urgently alternative forms of fertilizer and food packaging.

Reserves of land are unevenly distributed. In PR China and South Asia, which will support more than half the world's population by 2000, virtually all the potentially available land will have been put to use. In India, for example, already more than a third of the farming land is threatened with a total loss of topsoil.

We should bear in mind too that much of the spare land considered fit for farming is already playing a valuable role by, for example, regulating local climates, preserving natural drainage patterns and providing a refuge for the wild plant and animal species so necessary for maintaining industrial and agricultural production. Furthermore, these areas are the homelands of indigenous peoples – a neglected 4% of the world's population.

The vicious cycle

If we have learnt one very valuable lesson from the previous decade it is that environmental damage cannot be self-contained. The major cause of open tropical forest destruction is firewood collection. An FAO survey of 95 developing countries has found that now, and for the foreseeable future, 80% of the population in these countries will remain dependent on firewood. Already 1280 million people are experiencing a deficit, and that number is projected to rise to 3 billion by 2000. The tree cover is vital for maintaining soil fertility, keeping back the deserts and for preventing a desiccation of local climates.

The combined effect of agricultural stagnation, increase in human numbers, unfair land tenure, unemployment and population growth is continuously driving the rural poor onto marginal lands – hillsides, arid areas, rainforests – where farming cannot be sustained. Still others are driven into the already overcrowded slums and shanty towns.

The demand for cash crops has contributed to this vicious cycle. Some of the most fertile land in developing countries is given over to foreign exchange earning products, such as beef, peanuts, tea and coffee. Developing countries, in an effort to service debt repayments (US$93 billion in 1983) are turning unsuitable land over to cash crops. The price paid by the developed nations, relative to the cost of oil and manufactured goods, has fallen. For example, in 1980, profits from the export of one tonne of bananas were enough to buy only half the steel they would have bought 10 years earlier.

Many costly investments in ill planned irrigation systems needed to produce both exports and food consumption products have brought a disappointing return. After at first dramatically increasing production, some large-scale irrigation schemes have turned to a large extent into salty wastelands. A UNEP survey has found that world-wide some 40 million hectares are affected by salinization. Properly planned and well maintained irrigated agriculture is essential if we hope to feed a doubled population.

Top agricultural management priorities include: striking a balance between cash crop and food production for domestic consumption; concentration on increasing production in fertile areas; controlling demand from the affluent; improving natural resource-use efficiency; and improving energy, water and nutrient management. There is also vast scope for improving inequitable land tenure patterns which promote poverty, environmental degradation and inefficient natural resources use.

Though we have learnt much about human interaction with the environment, the uncertainties are many. It may well be that the carbon dioxide build-up will not affect the global climate, or that the protective ozone layer is in no danger or that the rates of tropical, moist forest destruction and rates of species extinction have been exaggerated. But these imponderables should not be used as an excuse for inaction. Nothing would please me more than to see UNEP's warnings of possible environmental disasters confounded. Yet, I must emphasize that there are dangers and opportunities and we must deal with both. Prevention is the only rational option.

Towards sustained growth

What the rapid increase in population has done, and will do, is to make sustain-able developments more elusive. This conclusion was reached in the most dramatic way by PR China. Its two child per family policy was replaced by a one child strategy when a projected increase of 400 million people was related to the availability of water, land, energy and other basic resources. PR China has achieved a reduction in the crude birth rate from 34 per thousand in the early 1970s to 20. The Chinese government concedes that, in a country where large families are highly valued, achieving this reduction has been a painful process. But it says that, given the assessment of its available resources, there has been no other option.

The confident expectation of the 1960s that when developing countries reaped the benefits of industrialization the rate of population increase would fall now seems like a forgotten dream. Some economists are now saying that the exceptional growth rates of the 1960s were an aberration. In the 1980s they say we must plan recovery on 1% and 2% economic growth. Low rates of population increase in the industrialized countries have helped them increase standards of living even in times of low economic growth.

We now know that the nature of the growth process is as important as the amount of growth. It must be measured in qualitative as well as quantitative terms. The risk of such an approach, of course, is that we may appear to be advocating a lower standard of living. But this need not be so. Sri Lanka, for example, has a per capita income of only US$350. In traditional, quantitative terms this is a poor performance. However, by placing a heavy emphasis on health, education and food production, Sri Lanka has achieved a life expectancy rate close to 70 years, a literacy rate of almost 90%, a reduction in the population growth rate in the 1970s from 2.4 to 1.7% and a 50% increase in per capita food production over the 1969–71 period.

The crucial importance of adequate rates of economic growth and of following the right development strategy is underlined by the plight of countries doing less well. For example, in Africa, parts of Asia and Central America hard won economic growth is being swamped by 2% and 3% annual population increases. Take food production: in Africa in the 1970s, food production actually increased by nearly 15%, but the rapid climb in population meant that it fell in per capita terms by more than 10%. For the first time ever, the World Bank has forecast a decline in income for 200 million people living in 24 countries of sub-Saharan Africa.

The increasing burden

There are a number of obvious ways in which the current rate of population increase will place additional burdens on the economies of many developing nations. For example, unemployment is already running high – over 20% in the case of Jamaica and Chile. According to the International Labour Organization, 75% more jobs will have to be created in Asia by 2000 to keep employment rates the same as in 1975, and more natural resources will be needed to keep those people employed. World-wide the demand for water needed by industry alone will increase over the 1975 level by a factor of 4.4. In Africa 350 million more school places will have to be created. Some time between now and the first decade of the next century more people will live in the towns than the countryside. In the developing world, 200 more cities will top the one million mark. Cities like Bombay and Mexico City will become megacities, with projected populations of 16 and 26 million respectively. Already services are inadequate to deal with the explosive population increase.

Between 1970 and 1980 the total number of people without access to safe water rose by a 100 million to 1320 million, and the proportion without decent sanitation increased from 25% to 27%. In Karachi, for example, one third of its 6 million inhabitants lack access to even the most rudimentary hygiene.

In sum, increasing human numbers means that each year the task of providing basic needs, while preserving the sustainability of the resource base, will become more and more problematical. The opportunity is to use family planning as a major contributor to economic and social development. For too long its role has been seen to be a passive one. It is time to change that.

The role of family planning

As the Bucharest conference recognized, access to family planning is a fundamental human right. The World Fertility Survey showed quite clearly that there is a massive untapped demand for birth control, not least because it lowers the death rate among mothers and their children. According to the World Bank, child spacing of at least two years can reduce child mortality by about 15%. It can also significantly reduce maternal mortality. The Fertility Survey found that about half of all married couples of reproductive age do not want any more children. Yet, mostly through unavailability of birth control, it is estimated that one third of all couples do not use any form of contraception.

More specifically, free access to family planning is a woman's right. The Fertility Survey showed clearly that in many societies women do not want to keep on having babies year in and year out. In many developing countries the main burden of firewood collecting, cooking and child rearing falls on women. With fewer births and less time spent collecting scarce firewood women would be left more time to devote to their own development and that of their children. I have in mind the Chipko villagers' tree protection and planting movement in the Himalayas, which began when the women took it in their own hands to save the trees. They saw more clearly than their menfolk how vital was the need to maintain the rapidly disappearing forest cover.

The reluctance on the part of some countries to pursue vigorously a population control policy perhaps stems, in part, from fear of a religious and political backlash. The cases where this has happened are few and far between. Perhaps one reason is that every single great religion endorses responsible parenthood. Another is that family planning is not an alien concept to the people of developing countries. In a major statement calling for a new commitment to population control, Mwai Kibaki, Kenya's Vice-President said, 'We must counter any African who says family planning programmes are foreign. Because the programme is African. In all our African tradition we were properly planned; my own father's family, for example, was beautifully spaced. The methods were different but they were African, and we always used to have them.'

To date 87 developing countries provide publicly sponsored family planning, according to the World Bank. A study by IPPF, the International Planned Parenthood Federation, found that no less than 138 governments support family planning, either directly or indirectly.

Developing countries with as varied socioeconomic systems as Cuba and South Korea and Mauritius and Mexico have introduced programmes well adapted to local and national needs. The hallmark of each programme has not been coercion but voluntary involvement. In India, the State of Kerala, for example, has logged remarkable progress in slowing the birth rate through a state-wide education and family planning programme. In the 1960s Indonesia launched a transmigration programme which was made almost redundant by a birth rate that stood at 3% in 1969. With strong government support, a programme pitched at the village level has reduced that rate by nearly a half in 1983. Villagers were given inducements to practice family planning in the form of new schools, irrigation pumps and so on.

There is overwhelming evidence to show that family planning programmes can create their own demand. In Thailand, for instance, where a major family planning programme has been implemented, in 1970 just 14% of the female population used some form of contraception. By 1981 the proportion had risen to 60%.

The dangerous opportunity

UNEP is looking to the 2nd World Population Conference to generate a new willingness to tackle our dangerous opportunity. UNEP's conviction is that family planning has an important role in safeguarding our human environment, but only when combined with economic and land reform, fairer terms of trade, conservation, and more emphasis on agro-forestry and the other prerequisites for vigorous and sustainable development.

Nor can maintaining and improving environmental quality be considered a matter for the rich alone. Poor communities, just as much as the relatively well off, want to maintain their forests, their wildlife and their unpolluted rivers and streams. There are few cultures which do not place a high value on preserving nature. It is a modern myth that the ethics of environmental protection are the concern only of the rich.

Uncontrolled population growth will do more than slow economic growth – it will quicken the disintegration of the traditional structure of societies. Long before we invented terms like 'carrying capacity', many societies lived in harmony with their environment. Development is not only about increasing material wealth, it is also about preserving our cultural inheritance.

Any new emphasis on family planning as a determinant in preserving our cultural diversity and in promoting sustainable resource use demands that governments and other decision-making bodies must plan for the long term. The task of this Conference is to produce an Operational Plan of Action; one that will both catalyse new action and sustain it. This is a formidable task in a world where few governments plan seriously beyond five years.

As I said before, a top priority for the 2nd World Population Conference must be to produce a Plan of Action which comprehensively reflects the complexity of the relations between people, resources, environment and development. I hope that the Conference, in addition to the recommendations already before you, will also recommend:

- first, a priority effort towards the identification of critical areas of ecological and social stress;
- second, the development and implementation of situation-specific population policies, in the context of the interrelationships between resources, environment, people and development;
- third, the strengthening of the interrelationships approach of the UN system as a whole through giving greater priority to the implementation of the Programme of Work on Interrelationships established by the General Assembly of the United Nations.

For my part, let me assure you that UNEP will do its best to assist with the implementation of these recommendations if adopted.

The spur to action is the absolute certainty that, unless we manage the world's finite resources more fairly and more wisely, political and economic chaos will follow. There can be no doubt that the sheer weight of human numbers, combined with resources mismanagement, will destroy the web of life which ultimately supports the economies of every UN member state.

19

Caring for lakes in a changing world

Statement to the Shiga Conference on Lakes

Otsu City, Japan, August 1984

A lake in this modern age is like a child in adolescence: part of a painful process of change, part of a complex web of troublesome relationships. Humankind's ties with lakes are the ties of parent and child; the child has a life of its own but it is indelibly imprinted with the influence and culture of the parent. In the same way, lakes, like children, are both givers of life and takers of life; capable of yielding great joy, capable of causing great pain.

This metaphor illustrates the fact that the relationships between humans and lakes are symbiotic in nature. Each influences the other, and the future of each depends on the other. These symbiotic relationships underscore the importance of our collective responsibilities to the freshwater systems of the world, of which lakes are an integral part.

Over 13 million people depend upon Lake Biwa and its watershed for drinking water, agriculure, industry and recreation. This lake, the largest in Japan and one of the oldest in the world with a history of some 2.5 million years, has always been a healthy source of fresh water. Yet over the past three decades, the lake has experienced a traumatic increase in pollution and a perilous slide in health.

Sadly this is very far from being an exception; in fact it is much closer to being the rule. World-wide the pressures on surface and groundwater supplies are increasing rapidly, exponentially in some cases.

In Japan, withdrawals for industrial purposes almost tripled in the 10 years ending in 1979. Thereafter, because of the widespread application of recycling, demand levelled out. Other countries, particularly in the developing world, are not so well advanced. The demand for water by industry alone is projected to increase by a factor of four between now and the year 2000. This will impose massive strains on already overstretched supplies. In West Asia some cities have been drawing heavily on fossil water supplies. In the drought-stricken African Sahel, women in the worst hit villages will queue all night to glean a few drops of water resulting from the nocturnal rise in the water table.

Demographic surveys carried out by the UN indicate that we must plan for the global population to level out at around 11 billion by 2150. Without a revolutionary improvement in the way we manage our water resources we will see an increasingly hungry world. The scenes of starvation and deprivation prompted by the recent droughts will be repeated many times over.

On a global basis, agriculture accounts for approximately 80% of water consumption. The most critical challenge will be to ensure an increased supply of water to irrigated agriculture. Many schemes are beset by salinization. In the Sahel, for example, as fast as new land goes under production, existing land is becoming useless. Until recently, Pakistan was losing over 240 000 hectares of fertile cropland every year to salinization, and nearly 10% of the total farming area in Peru is affected by salinization. In a recently completed UNEP survey we found no reason to believe that this situation would change in the near future.

In the semi-arid areas in particular, groundwater abstraction – primarily for irrigation – far exceeds the rate of recharge. The result can be salt water intrusion, land subsidence and lowering lake levels. The upshot is to increase dramatically the cost of pumping for farming, domestic and industrial needs. In the Indian State of Tamil Nadu the water table in 37 observed wells dropped by over six metres over a six-year period in the late 1970s.

I cite these facts and figures to indicate the growing demands on our water resources. Two years ago, at the commemorative meeting in Nairobi of the Stockholm conference, I predicted that if energy was the resource issue that attracted more attention in the 1970s, as the 1980s unfold water will take its place alongside energy. There are very clear indications that this is happening: this Conference is an example among many of the growing concern for our threatened freshwater supplies.

Pollution of water systems

While overusing water resources because underdevelopment forces us to do so may be understandable, I fail to understand why human beings continue to pollute our precious water-systems needlessly. We abuse and often destroy the very environmental systems upon which our well-being depends. The near destruction of Lake Erie in North America by industrial pollution, the destruction of over 20 000 lakes in Scandinavia and Canada by acid rain, the siltation and eutrophication of thousands of lakes in developing countries by imprudent development and sanitation practices; all of these merely hint at the extent of the problem.

The degradation may even be deliberate. Lakes near industrial parks may be consciously sacrificed in the name of economic development. Traditional cost–benefit judgements may be unwise, as in decisions to allow discharges of non-degradable toxic chemicals to avoid the expense of pollution control. Sometimes, people just do not care.

But these are the exceptions. More frequently, lakes degrade because we do not fully understand the limits of resilience of our natural systems. This is especially true in the developing countries in general, and in the tropics in particular. I believe that the real reason behind this degradation is that too many people have not yet realized the critical linkages between development activities and the environment.

Relationship between environment and development

Our understanding of the relationship between environment and development has undergone a profound change over the past 10–15 years. At the end of the 1960s, almost everyone was convinced that you either had one or the other; that is, if you wanted to have development then the price to pay would be a loss in environmental quality. That view is fortunately changing as we have come to realize the complex nature of environmental problems and concerns. Environmental problems and concerns are, by definition, holistic and integrated. They cannot be put into compartments. They are linked one to the other and any attempt to put them into sectors or consider them separately leads to frustration.

This holds good as much for economic and social processes as for physical relationships. We all know the results of sectoral planning when we have tried to deal with different sectors, like agriculture and industrial development or infrastructure, without considering the parameters and constraints of the economy as a whole. The results over the last 35 years have not been very positive. Today there is, accordingly, a search for integrated physical, socioeconomic and environmental planning – a search for strategies that will provide a minimum environment with sustainable development. The words minimum and sustainable require explanation. By minimum environment we mean a situation in which we are not looking for 100% pure air or 100% clean water, or leaving the national resources completely untouched. We are talking in terms of trade-offs and mixes. For sustainable development, what we have in mind are not short-term, one-time projects but growth strategies which can be sustained over a long period of time. This means that the availability of natural resources, both renewable and non-renewable, and the regenerative capacities of the environment are kept centrally in mind, permitting that kind of development.

Another aspect of environmental problems and concerns to which attention must be drawn is that causes and effects of these problems are complex and sometimes unknown. Most of the effects seem to be synergistic in nature; they could also often be irreversible.

Prevention better than a cure

Two conclusions follow for practical policy. One is that, in most cases, it is more economical to treat environmental problems at an early rather than at later stage because costs of correction increase in an exponential manner. This has been well documented in the literature on the cost–benefit analysis of environmental protection measures. At the Conference on Environment and Development held by the OECD in June this year, the perception that received common support was that it was both more efficient and effective to move away from react and cure to anticipatory and preventive policies.

Economic justification for conservation

The other conclusion is economic justification for conservation measures. As you know, the IUCN, in co-operation with UNEP and others, launched four years ago the World Conservation Strategy, which, in turn, has encouraged a number of countries to introduce national conservation strategies. Although conservation strategies are supported intuitively and sometimes on the basis of empirical evidence, many international development financing institutions, both bilateral and multilateral, would feel more comfortable if they could be provided with economic justification for conservation as a practical policy alternative. When the results of present action could be irreversible – as is the case with the loss of genetic resources, tropical forests or soil fertility, it can easily be proved that conservation pays.

Sound environmental management

Most industrialized countries have learnt from experience that in order to have long-term sustained development it is necessary to have sound environmental management. The developing countries are only gradually coming to accept this view. On the other hand, the developed countries have yet to accept fully the second basic principle in the interrelationship between environment and development, namely, that economic problems create environmental damage, which in turn makes structural and economic reforms more difficult. If this was accepted, then agreement would have emerged that the present economic/financial crisis cannot be cured, except with new, less wasteful and environmentally sound resource-use patterns.

Two examples stand out. First is the arms race. The world is spending billions of dollars in producing arms which not only poison the environment but use up scarce resources. One has only to look at the figures from 1972 to today and compare them with the health expenditures, especially in developing countries. The other example is the debt burden of the developing countries. Its interest now stands at US$93 billion. In the repayment of this debt the developing countries are using much more than their economic surplus. They are making extremely large demands on their natural resources. Such demands cannot be sustained and could lead in the long term to serious ecological imbalances for the whole world.

A clear and comprehensive appreciation of the nature of environmental problems leads to an understanding of the interaction between man and nature on the one hand, and the design of practical development policies and objectives on the other. Interactions among forces of change, namely resources, environment, people and development, make it necessary for governments and decision makers everywhere to think in terms of trade-offs between alternative courses of action. Each possible development objective must be carefully evaluated in order to identify its environmental consequences.

It is important to avoid harmful consequences inflicted by an action in one sector upon other sectors. For example, say an energy department builds a hydroelectric dam, but the forestry department does not follow a sound watershed conservation policy. The result will be a silting up of the dam and a failure to meet energy production targets. A 1979 report by the United States Agency for International Development (USAID) observes that deforestation in northern Luzon in the Philippines has silted up the reservoir of the Ambuklao Dam so fast that its useful life has been reduced from 60 to 32 years. This is only one example; there are many more.

Need for an integrated approach to environmental protection

The recognition of the critical relationship between environment and development leads us directly to the need for new and more integrated approaches to environmental protection. A common philosophy must underlie our management efforts, which I would sum up as follows:

1. Economic and social development must be pursued to meet the basic human needs of all people and to secure a better future for them.
2. Environmental processes must be thoroughly and widely understood.
3. The productive capacity of the environment must be maintained and resources used rationally.

The management of lakes presents an opportunity for turning the theories into action. When the catchment system is considered as a whole, the lake can be seen as a largely self-contained entity, with physical processes, inputs and outputs which can be described in fairly precise terms. The simulation models I have seen – those describing eutrophication, distribution and dispersion of pollutants, production of fish, and the phosphorus cycle, to name a few – are among the most well developed models in any field of the environment.

In the temperate regions of the world, including Japan, our understanding of limnology has developed to the point where we can think in realistic terms about managing lakes and their catchment systems in an environmentally responsible way. Development plans are well laid and programmes are already underway. The environmental dimension must be added, and added soon.

The greatest challenge is caring for lakes in the developing world where financial resources and expertise are scarce. New and more integrated management approaches are needed if we are to meet this challenge. Drawing on her own experiences in water reuse, there is much Japan could do to assist its neighbours.

Effective lake management depends upon how well the lake system is managed as a whole. A lake cannot be divorced from its watershed, from its rivers, from its land or from its people. Effective management must integrate all these aspects. In the same way, all major environmental functions of a lake must be considered simultaneously, including the natural resources, the local ecosystem and the landscape. The current practice is that only the first of these environmental functions – the natural resources – is considered in the course of socioeconomic development. This single function is the normal driving force behind urbanization, industrialization, agricultural development, and so on.

The result is often eutrophication, deterioration of water quality, drastic changes in fauna and flora and the spread of water-borne diseases. The debilitation caused by contaminated water can be a severe brake on development. A recent survey in South America found that the provision of clean water can recoup costs up to 7 times over as a result of the number of working days saved. People and their relationships to local ecosystems and landscape must be brought into the picture. Possible environmental changes can thus be balanced consciously against development needs; lakes and their watersheds can be considered as an integrated whole; and management can be better directed to our long-term self interests.

Realizing the value of natural resources

We need to look upon lakes and the other component parts of freshwater systems as resources every bit as valuable as, say, industrial plant, foreign exchange reserves and the other traditional measurements of wealth. Our global imperative is to reform the neoclassical approach to national accounting to take in long-term benefits, health, spiritual, aesthetic and other factors. UNEP, in cooperation with other UN bodies such as the World Bank, is carrying out pioneering work in this area.

I do not underestimate the difficulties. Such an approach requires trade-offs and short-term sacrifices. Let me give you an example. In the Brazilian Amazon forest, each year the great Amazon river and many of its enormous tributaries flood vast areas of the forest, creating one of the world's largest freshwater lakes. This floodplain forest, called the Varzeas, is the most robust one of the Amazon ecosystem, suitable for rice growing and water buffalos. But the Varzeas are a unique and productive ecosystem, supporting the world's largest inland fishery. The neoclassical approach would be to liquidate this natural capital and convert to rice and other forms of cultivation. But in the process, the floodplain forest, the fishery, the range of flora and fauna and the indigenous people would be severely disrupted. Surveys carried out by several research institutes indicate that, if these social and environmental factors are taken into account, the most sensible policy, in the long term, would be to exploit the forest intact. These are the kinds of development decisions that confront ill equipped developing countries on a day-to-day basis.

Environmental management of lakes

Developing water systems involves a complex range of different types of decision making. In every case, lakes included, a precondition is a thorough understanding of how the ecosystem functions. In specific terms, there are three terrestrial elements of lakes; the water body, the shoreline of the lake including the inner watershed, and the outer watershed. While all are interconnected, each element can be described in different environmental terms and fully integrated decisions can be made accordingly. Similarly, the underlying geological structures and groundwater conditions can be taken into account.

Conceived in these terms, an integrated series of concrete management steps can be taken to merge the environmental aspects of a lake with its development. These steps include:

1. suitable monitoring and evaluation of the lake environment;
2. effective management of water quality, including treatment of municipal and industrial effluents and proper solid waste disposal;
3. effective control of water quantity, including sound management of flood plains;
4. controlled use of pesticides and chemical fertilizers, considering the overall needs and resilience of the entire watershed;
5. improved agricultural and silvicultural practices to improve protection of catchment systems;
6. proper land management to prevent erosion, conserve topsoil and direct urban growth;
7. enlightened conservation programmes to avoid drastic changes to the flora and fauna of the lake.

These concrete management steps are not one-off actions to be taken in a vacuum: they are part of a dynamic decision-making process. And we have the tools we need to make this process work. Environmental impact assessments and cost–benefit analyses of environmental protection measures have proven to be practical in use and effective in achieving environmental results. These tools are gaining acceptance among environmental managers in developed and developing countries alike, and they hold great promise for improving our management of lakes.

By working within this management framework and by consistently considering the environment over time, the results are dramatic: planning improves, decisions become more sensible, and the long-term development interests of local population are preserved. One caution, however: this framework only works when suitable legal and institutional controls exist. Lake authorities and commissions have proven effective; other arrangements are possible. But management decisions must be put into effect to have meaning. Enforcement powers may be necessary.

All too often action is taken only when the lake has already been damaged. Worse, I have noticed that a large number of developing countries are focusing solely on industrialization and economic development, and still do not see the crucial long-term value of protecting their lakes and watersheds.

Governments do not make decisions in a vacuum. They do so through support or pressure exerted by citizens, and such support and pressure will be effective only if people, first, understand more about the environment, and, second, feel that they have a positive role to play in decision making at all levels. In Australia public concern over a proposed dam in an unspoilt catchment area in Tasmania was effectively mobilized and the scheme abandoned by a new administration which had benefited from the public's concern.

If lakes and other natural resources are not receiving enough attention in the development decisions made by governments, then that is a reflection of a failure by environmentalists to make their existence and value known. Greater understanding must be achieved, locally, nationally and globally.

Thus, the ultimate value of this conference may lie in an increased public awareness. Your excellent scientific work and the resulting improvements in our understandings of lakes are the foundations upon which we can build. As I look across this room, filled with distinguished representatives from all over the globe, I take heart. With each of you enlisted in the common cause, perhaps we are in fact moving closer to a broad-based awareness of the importance of lakes. From this awareness, we may be able to achieve a truly global consensus on the importance of their protection. We at UNEP are vitally concerned with achieving this global understanding of lakes and their watersheds, and with translating this understanding into action. I expect the worst if we do not succeed.

Potential conflict as a result of water shortages

National and global security is at stake. Shortages of fresh water worsen economic and political differences among countries and contribute to increasingly unstable perceptions of national security. On a global basis there are 214 river or lake basins shared by two or more countries. During the last 15 years conflicts have emerged over the development of international rivers, such as the Colorado involving the USA and Mexico, the Euphrates between Iraq and Syria, the Indus involving India and Pakistan, the Jordan river involving Israel and Jordan and the La Plata between Brazil and Argentina. Increasing populations and rising demands will increase the scope for conflict. For example, the Ganges Basin alone may have to support some 500 million people by the year 2000.

There is an urgent need to identify existing and potential conflicts and to develop guidelines and regional machineries for resolving them. The UN could play an important role in such a development. This conference might wish to make such a recommendation. If one were forthcoming, UNEP would undertake to cooperate in its implementation with relevant organs of the UN system.

Action to clean up lakes

Each of us shares in the danger and each of us shares a responsibility to act. The good news is that people all over the world are beginning to take action, and that many lakes are now healthier as a result. An example is Lake Biwa, which has suffered terribly from over-rapid agricultural and industrial development. Prompted by a strong public outcry – an all-too-frequent prerequisite for action – a broad-fronted clean-up programme began in July 1980. But the importance of the efforts at Lake Biwa goes beyond the remedial steps. A comprehensive, forward-looking management programme has been initiated to avoid similar problems for future generations – a tribute to the enlightened leadership of Shiga Prefecture.

Other examples abound, including Lake Balaton in Europe, Lake Washington in North America, and Lake Victoria in Africa. I regard these as happy signs that environmentalists and governments all over the world are beginning to realize their common problems, and to take appropriate action.

UNEP's role is to prompt this action. By the terms of our charter, UNEP is a catalytic organization. This means we must carefully choose where and how to intervene in issues which involve the environment. Within our catalytic role, we are now working actively to finalize and launch a comprehensive programme on the environmentally sound management of fresh-water bodies and their basins. This new programme is devoted to rivers and underground water aquifers. Our efforts will be directed along two tracks: first, helping to fill in critical information gaps; and, second, helping to put environmentally sound management principles into practice.

Tropical lakes

The most critical information gaps we face relate to tropical lakes, ie lakes in a large number of the world's developing countries. Pilot projects have been proposed at three tropical lakes: Lake Managua in Nicaragua, Laguna de Bay in the Philippines, and the Nyanza Gulf region of Lake Victoria in Kenya. Each pilot project would identify specific information gaps, transfer knowledge from developed countries to developing countries, and train scientists and administrators in tropical limnology and tropical lake management.

Pilot lakes and guidelines for lake management

The second thrust of our effort is to develop a world-wide network of well managed lakes and watersheds. To set the stage, we are now working on a set of comprehensive guidelines for environmentally sound lake management, which should be available in 1986. To put these guidelines into practice, we are planning a system of pilot lakes, both man-made and natural, which can be used as demonstration projects. We are suggesting Lake Tanganyika as an appropriate starting point.

The larger message is this: the fine work of this conference, and the dedication of the limnologists and watershed managers of the world, will amount to nothing unless they are ultimately translated into hard, concrete actions at individual lakes.

I thus applaud the efforts of the Shiga Prefecture and the people of Japan to clean up Lake Biwa and to establish policies for developing the lake's resources in an environmentally sound way. I applaud the many similar efforts in other parts of the world. But these efforts are starting points, not end points; as the pressure mounts so we need to increase our efforts before we achieve the transition from isolated pilot projects to world-wide mobilization.

Nationally adapted and locally appropriate policies and guidelines for the environmentally sound management of lakes are needed everywhere, preferably as integral parts of national development plans or conservation strategies. UNEP's comprehensive guidelines will hopefully form a relevant working framework, but each country, and each locality, will have to take cognizance of its own development needs and consider the environment in its own way.

Citizen involvement

Local, citizen-backed action needs to be encouraged. To do so, we must widen our dialogues and carry our message to all people involved in the development of lakes and their basins, including industrialists, farmers and city planners. Citizens must be involved at every level. Thus, public awareness campaigns must continue to occupy a prominent position in our efforts, backed by well considered outreach programmes and supported by decision makers from every sector.

International cooperation

To achieve world-wide mobilization, we must achieve greater international co-operation. International lakes have sometimes served as vehicles for cooperation among neighbours, but much more is possible and necessary. We in UNEP have gained very good experience from the Regional Seas Programme, whereby numerous governments, often including traditional adversaries, sign agreements and develop specific action plans to protect their common interest, the regional sea. The same approach seems to be perfectly possible, and entirely appropriate, for freshwater systems.

Agreements to protect freshwater systems

Thus, I propose a system of agreements among governments to protect major international freshwater systems, including man-made and freshwater lakes. Such agreements would prescribe diagnostic assessments and monitoring; develop specific action plans for environmentally sound management of the water resources; anticipate and resolve potential conflicts and establish small, institutional bodies to coordinate and oversee the implementing steps taken by each member government. UNEP is now working to create such a regional programme in Africa, for the Zambezi River and Lake Tanganyika. Other major regional freshwater systems of the world, touching on developed and developing countries alike, hold similar promise as vehicles for closer international cooperation and action.

The future of the world's lakes

At the beginning of these remarks, I painted a picture of lakes as adolescent children, painfully shedding the innocence of youth as they experience the strains of modern-day development and economic growth. Our relationships with these adolescent children are essentially human: their identities and futures largely reflect our own.

The processes I have described, including the concept of environmentally sound development, are designed to restore a healthy, nurturing relationship between parent and child, between people and lakes. The Earth's resources are not inexhaustible, and the time is not far off when our development practices must reach a sustainable, steady state if we are to survive and prosper as a species. Freshwater lakes are a vital part of our natural resource heritage. Properly cared for, lakes will help support humankind indefinitely. But if lakes are abused beyond their limits, the parent, like the child, will die.

Each of you has a part to play in determining the future of the world's lakes. Each of you has an important contribution to make here, and a vitally important role to play when you return to your countries.

I am proposing that we establish an international Shiga committee to follow up on the work already done here. With a focal point established in this prefecture, you should try to meet annually, or biennially, at the site of another lake where steps have been taken to safeguard its environmental health. In this way we could sustain the spirit of Shiga and draw public attention to the plight of lakes and their nurturing river systems.

20

The global dimension to conservation issues

Address to the National Conference on Economic and Environmental
Policies for Sustainable Development

Sydney, Australia, September 1984

Until recently, environment and development were viewed by many as opposites –
you either had one or the other. The word 'conservation' was taken literally to mean
no change at all, and the single-minded pursuit of short-term profits was seen to
preclude environmental protection.

The terms 'ecodevelopment' and 'environmentally sound development' came into
popular use more than 10 years ago, sounding a hopeful and constructive note. Yet,
environmentalists and industrialists continued for some time to view each other as
adversaries. Perceptions were polarized.

But now, through a decade of global conferences, regional dialogues, and hard
work everywhere to broaden discussions and awareness, a global and much more
integrated perspective is emerging. Nowadays we rarely argue over whether
environment and development are mutually exclusive; we mostly argue over the
most economical ways to achieve a joining of the two. Our conversations are no
longer so one-sided. The balance among the leaders I see before me today – environmentalists, businessmen, government managers and citizens – is yet another sign
that the dialogue on the environment is broadening and deepening.

This conference, significant as it is, is not an isolated event. It is part of a world-wide
awakening of leaders in every sector to the importance of the environmental
dimension of economic growth and development. Nineteen eighty-four has seen
major international conferences in Munich, Washington, Moscow, Buenos Aires,
Japan, and now Sydney – the issue is truly coming alive. In November of this year,
in Versailles, industry leaders from all over the world will meet with governments,
NGOs, labour parliamentarians and scientists to discuss environmental management.
Never before have I witnessed such a blossoming of interest.

Development, population, resources, global security – we now perceive an
environmental dimension to each. While we previously thought that these four
major world processes were separate and distinct, we now know that all are
interrelated, and that the environment is what ties them together.

In the 1970s the emphasis was largely upon the negative environmental aspects of
development: pollution, destruction of habitats, and so on. And the negative
attention was not only localized. These negative environmental aspects of development, combined with other factors, often presented challenges of a global dimension:
acid rain and possible climatic changes resulting from the combustion of fossil fuels
to meet energy needs, to name just two examples.

Perhaps the most significant understanding now is that the absence of development
can also be a major source of environmental problems. Poverty, the root of so many

of the world's ills, breeds environmental destruction. The struggle merely to survive forces people onto marginal lands, to overburden exhausted land, to destroy the tree cover. These pressures inevitably lead to destruction of natural resources. The environmental manifestation – deforestation, desertification, reduced fertility of soils, disease – occur on a local scale and merge to become global problems. The refugee camps, food hand-out centres, slums and shanty towns are the end of the line. These manifestations are infinitely more serious than any arising from industrialization or overconsumption.

Sustainable development

Environment and development are complementary. Without environmental protection, development cannot be sustained. Without development, poverty cannot be erased and the quality of life cannot be improved. In its broadest terms, the concept of sustainable development encompasses:

1. help for the very poor, because they are left with no option other than to destroy their own environment;
2. the idea of self-reliant development, ie within natural resource constraints;
3. the idea of cost-effective development, using non-traditional economic criteria;
4. the perception that development should not degrade the environment nor result in reduced productivity in the long run;
5. the great issues of health control, appropriate technologies, self reliance in food, clean water and shelter for all;
6. concentrating on the environmental systems at greatest risk, be they arid lands, watersheds, moist forests, or areas of rapid expansion.

One result of the debates of the 1970s is that in the 1980s a consensus is emerging on the means to achieve this concept of sustainable development. These means include:

1. improving indigenous capabilities for managing the environment and natural resources;
2. building upon past experiences to ensure that we all learn from past mistakes;
3. ensuring that environmental considerations are built into development planning as a routine matter;
4. gathering hard environmental data to serve as a basis for sound development planning;
5. informing the public of what is at stake;
6. concentrating on the environmental systems at greatest risk, be they arid lands, watersheds, moist forests, or areas of rapid urban expansion.

Economic justification for conservation

When it comes to applying the elements of sustainable development, we encounter major constraints. UNEP's work on environmental cost–benefit analysis has helped to overturn the idea that protection of the environment involves additional costs with no return. Numerous studies have shown that recycling wastes which would otherwise become pollutants can boost profits equivalent, for example, in the case of the US 3M Corporation, to $200 million in additional annual sales turnover.

But, in general, neoclassical economics and environmental management are difficult to merge, mainly because environmental values are difficult to quantify. Environmental consequences such as air quality, soil fertility and so on have important implications for human health or agricultural productivity but, although these may be important economic factors, they are not amenable to exact measurement. How, for example, do you put a price tag on keeping a river unpolluted or on preserving an endangered species that may turn out to be of major economic value? What is the economic justification for conserving slow-reproducing renewable natural resources such as whales, redwoods or tropical hardwoods? The classic economic approach would be to destroy them completely and invest the profits in a more lucrative enterprise.

However, when we see how marine pollution and destruction of fish breeding grounds is undermining marine-based economies; when we see how deforestation is affecting local climate patterns; and when we see soil erosion undermining agricultural productivity, we see also the bankruptcy, quite literally, of the traditional approach which measures successful development purely in high GNP growth rates. Such an approach creates powerful incentives for countries to 'liquidate' their natural resource base as rapidly as possible.

What needs to be achieved is an accommodation in economic decision making of the long-term social, aesthetic, spiritual and, of course, economic consequences of the irreversible liquidation of natural capital. It is a matter of keeping the options open for ourselves and for future generations. In short, the need is to define the specific principles of sustainable development and couple them with current economic theories and practices.

Natural resource accounting systems

This demands a re-evaluation of accounting systems. All too often the natural patrimony is discounted; the result is that resources not accounted for are apt to be wasted or mismanaged. In UNEP we have embarked on a programme aimed at producing modified accounting systems that would assist governments in improving resource management. The aim is to develop methods that internalize natural resources stocks and environmental services in national accounts.

Risk management

Another major element in our considerations of sustainable development is the fact that the 'bottom line' in environmental management is risk management. The future is filled with risk and uncertainty and nowhere is uncertainty so pervasive as in everything to do with environmental and resource conservation problems. The causes and effects of these problems are complex and largely unknown. The effects are frequently synergistic in nature and sometimes irreversible. As a result, we simply do not have enough information or the analytical apparatus to predict the long-term consequences of interventions in the environment.

For example, we do not have adequate information, nor have we classified and categorized the genetic resources of the world's tropical forests, the home of at least 40% of all the plant and animal species on earth. Yet each year we lose an area of tropical forest twice the size of Switzerland. At the very minimum, we will lose one eighth and probably more of the existing area by the turn of the century. Up to a million species could be lost.

We have no way of telling how many potentially valuable species are being wiped out. An even greater threat is posed by the phasing out of traditional crops, 'cultivars'. Greece, for example, has lost 80% of its traditional wheat strains since the second world war. A parallel trend is the redundancy of the primitive ancestors of domesticated livestock. The world is becoming dangerously dependent on a few high-yielding species.

The science of bioengineering is advancing rapidly. A recent report to the US Congress predicts it will be the last great revolution of this century. But bioengineers now, and in the foreseeable future, will depend on nature's storehouse. *In situ* conservation is still our best safeguard for the future.

Thus, the economic justification for conservation lies in the fact that there is a positive value in retaining an option to preserve or develop in the future. This is of course dependent upon a number of factors but could be calculated using available economic techniques.

Role of UNEP

By the terms of our charter, UNEP is not an operational agency. UNEP was created to act as a catalyst, coordinator and stimulator of environmental action. By spreading the environmental message, by stimulating others to action and by intervening on behalf of the environment at opportune moments, UNEP seeks to persuade others to do the work it cannot do alone. Sparking environmental awareness on a global scale is one challenge, but achieving coordinated reponse among nations is even more difficult.

Regional Seas Programme

In the mid-1970s, UNEP pioneered a series of agreements among groups of nations to clean up and protect their regional seas. These seas include politically complex regions such as the Mediterranean, the Caribbean, and the Arabian/Persian Gulf. This programme has been remarkably successful. Even traditional or present day adversaries – Iran and Iraq, the USA and Cuba, Israel and Libya – have joined together in common efforts to protect their common resource.

These examples are conspicuous proof that, beneath national pride and outward politics, the nations and peoples of the world treasure their environment and are willing to act to protect it. Where public and political wills exist, the goal of development without destruction is attainable.

Resources of Antarctica and the Southern Ocean

An important test of nations' willingness to safeguard their environment will be the conservation of the living and non-renewable resources of the continent of Antarctica and the surrounding Southern Ocean. With the notable exception of the great whales, Antarctica's as yet undetermined treasure of resources remains untouched. But distance and the inhospitable climate that have kept them beyond our reach are becoming less formidable obstacles to a resource-hungry and innovative world. Krill fishing has begun and mineral exploration has been carried out. But we know almost nothing about the workings of the Antarctic ecosystem. Much more research needs to be carried out before we can contemplate fixing a meaningful sustainable yield and establishing environmental safeguards. The Antarctic Treaty powers have kept Antarctica weapons-free and a continent reserved for peaceful scientific investigation. But the Treaty expires in 1992 and developing countries have seriously questioned the right of predominantly developed countries to use Antarctica for their own purposes. The world's public will be looking to the international community to establish not only a sustainable but also an equitable regime for this last untouched part of our global environmental heritage.

Increasing environmental awareness

There is evidence that the necessary public and political wills are emerging. The public information campaigns of the last decade, the increasing awareness of government leaders, businessmen, journalists and citizens have all had their effect. Leaders of the last administration in Australia felt the full power of public concern, aroused over the proposed dam on the Fraser/Darling catchment in Tasmania.

And now with the coming into production of the new gold mine on the Ok Tedi river in Papua New Guinea near the Indonesian border, run by a business consortium in which Australian business is a partner, we read and hear of fears about its potential environmental effects. Already international concern, particularly in Australia, has been expressed over a series of serious cyanide spillages.

In recent years Australia has taken a keen interest in conservation measures both at home and abroad. This is the result of a correct appreciation that resources conservation and environmental concerns are interdependent and heedless of political boundaries. No country can isolate itself from the rest of the world, either morally, strategically, economically or, more fundamentally, environmentally. We inhabit the same earth and we are tied to it for our happiness and well-being.

Cooperation and collaboration

Australia's continuing support for this approach is important and any strengthening of effort would be widely appreciated in the internatonal environmental community. Australia may also find that close collaboration with other bilateral environmental and aid agencies could yield positive results. It can share with them its own experiences in integrating environmental concerns in development activities – the bottle necks encountered, the breakthroughs achieved, the new methodologies developed – and learn from theirs. In this connection, I would like to mention the Committee of International Development Institutions for the Environment (CIDIE). The CIDIE came into existence when nine major multilateral development institutions – the World Bank, the EEC, the UNDP, the Asian Development Bank, the Caribbean Development Bank, the Organization of American States, the African Development Bank, the Inter-American Development Bank, and the Arab Bank for African Development – signed with UNEP in 1980 a Declaration of Principles. They have since been joined by the European Investment Bank. CIDIE will welcome the participation of bilateral aid agencies as full members or their attendance at its meetings.

The scope for mutually beneficial cooperation is vast, and there are some encouraging signs. One example concerns genetic conservation; wild legumes from Mexico were bred into a new crop in Australia and this crop is now widely grown in Third World tropical countries. We can be encouraged too by the increasing importance developing nations now attach to preserving their wild plant and animal resources. Increasingly, protected areas are being included in this new 'integrated' approach to development. A paper to the recent World Parks conference in Bali listed no less than 120 success stories along these lines. Australia may wish to cooperate with its neighbours in similar activities. One such area where Australia has a wealth of accumulated experience is desertification control.

Desertification is a major and growing menace at the global level: 135 million people in rural areas are now known to be severely affected by desertification, compared with 57 million in 1977. The ill effects of desertification and other environmental problems are always felt most acutely by local people. But our challenge is global. Since UNEP's inception in 1972 we have been trying to persuade governments to 'think globally and act locally'. Our job at UNEP is first, to sustain and enhance an acute public and political understanding on a global scale, and second, to channel effectively the force that these wills represent.

The need for action

Our efforts to channel this message and the public and political forces behind it depend not so much on us, but on you. What we need now is hard, concrete action to multiply past, positive experiences, and to deepen our understanding of sustainable development and to apply it on a new and far wider scale.

Applying these hard, concrete actions depends on you. Despite the recent economic recession shared by so many countries in the world, Australia stands at the edge of a period of unprecedented growth and development. Its natural resources are vast, its land seemingly endless. But every single act of development, every increment of growth, carries a potential environmental price tag. A multitude of development alternatives are in front of you, some of which lead to long-term, environmentally stable development, and some of which do not. The end of the 1970s was marked by a negative attitude among many that development was bounded by what were often termed 'limits to growth'. We must live within the limits of our means, it is true, but through well managed environmental development we can improve our lot dramatically.

Environmentally sound development means that it is much less costly and much more efficient to take environmental considerations into account from the beginning rather than at the end of decision making and planning for development. At one end is the case of retro-fitting of pollution control equipment which is demonstrably more expensive than built-in devices. But even in cases where it is not simply a question of engineering design, there is a strong case, in terms of economic incentives, to move away from react and cure to anticipate and prevent actions. Damage costs increase in an exponential manner and often quadratic functions have to be considered. In this situation, control costs are less complicated, more manageable and less costly at the beginning of an intervention in nature before temporal and spatial considerations and consequences aggravate. There is a mountain of evidence to support this proposition.

The way ahead

In May of this year, a conference was held near Washington, DC to consider the challenges of the next century. The meeting, entitled 'Global Possible' was attended by over 75 world leaders in government, industry, science, agriculture, energy and the environment. The conclusion of the group was unabashedly optimistic: current negative environmental trends can be reversed, and we can build a sustainable society. I share their optimism, for in the last analysis, building a sustainable society is a matter of common sense and of benign self-interest.

21

Energy use and environment in the Asia/Pacific region

Keynote Address to the International Petro-Pacific '84 Congress

Melbourne, Australia, September 1984

The Australian Institute of Petroleum is to be highly commended for organizing this timely Congress. It is a sign of a new awareness among governments and businesses of the importance they attach to the environment. The time has passed when environment was seen as a barrier to economic development. The fact that the environment forms an integral part of your agenda is a positive statement in itself.

Our overriding concern in UNEP is to promote the concept that economic growth can neither be equitable nor sustainable unless environmental considerations are taken into account. This means we take the broad view of the term 'environment'. Pollution control and conservation of local resources and ecosystems are important environmental concerns. But the fundamental environmental concern is the present quality of our lives, and the prospects for our future. We thus perceive all dynamic processes of human society as environmental in nature: population growth, economic development, health, resource use and regeneration, national and global security, and, of course, energy consumption.

UNEP is unreservedly pro-development. People are the most important resource of this planet, and development is necessary to improve the quality of lives of people. But to benefit people truly, the development process must be sustainable and to be sustainable, the regenerative capacity of nature and the resources of our physical world must be used wisely.

Our new understanding of the interrelatedness of environment and development is the most important perception of the previous decade or so. There are no frontiers between the two. In my address I will stress the interconnections between environment and energy utilization in the overall process of social and economic development. I will try to indicate that, not only are environmental objectives compatible with those of development, but also that the overall target of achieving sustainable economic growth will not be reached unless a reconciliation is realized.

The United Nations Environment Programme

But before I move on to the main part of my address I should, perhaps, say a few words about the United Nations Environment Programme (UNEP). We are a small organization with headquarters in Nairobi. Unlike the big agencies such as WHO or FAO, we are not an operational agency carrying out projects in the field. Our main job is to persuade or 'catalyse' other UN agencies, governments and increasingly business and industry to take the environment into account in major development decisions.

Within the limits of our restricted financial and manpower resources, we have registered reasonable successes. This has been made possible by a consistent willingness on the part of governments to put the environment beyond the kind of political differences which hamper positive action in other spheres. A case in point is our Regional Seas Programme which now involves over 120 states world-wide. In the southern hemisphere we have launched the contiguous East Asian Seas and South Pacific programmes. Both have succeeded in involving the states of the two regions in a cooperative endeavour to safeguard their marine and coastal environments from oil pollution and other threats.

I am particularly pleased to note that UNEP has played a leading role in bringing down the barriers between the business and environmental communities. In the mid-1970s we set up our Industry and Environment Office in Paris which has been our focal point for spreading the dialogue with industry, a process to which the petroleum industry has extended its cooperation.

In France in November of this year, a number of leaders of different industries will convene, in cooperation with UNEP and the ICC, the World Industry Conference on Environmental Management to discuss with governments, parlimentarians, labour, NGOs and the scientific community, the role of industry in environmental management. The scope of the conference and the breadth of participation – over 300 participants from more than 100 countries – testifies to the growing under-standing among industrial and business leaders that environmental management is crucial to the responsible conduct of their businesses.

Energy use in the Asia/Pacific region

The Asia/Pacific region is one of stark cultural and economic contrasts. On the one hand there are industrialized nations whose economies are largely dependent on fossil fuels, on the other developing nations whose rural majorities depend on fuelwood and dung, and in the middle the industrializing nations. No single energy strategy can match the needs of this polyglot region which is home to more than half the world's population. But in all cases the threat to the environment is great: to the better-off it comes from oil pollution, carbon monoxide from traffic, gaseous emissions from factories and power stations; to the poor it comes from deforestation which is a major cause of soil degradation, desertification and plant and animal extinctions.

The potential energy requirements of the region's developing countries are huge and if they follow the old development and consumption patterns, which were neither efficient nor clean, then the environmental consequences will be severe.

However, there should be no need for this. During the past two decades, there has been a significant increase in knowledge on designing, constructing and operating energy efficient systems in all sectors of application. Similarly, measures to mitigate environmental consequences from the production and use of energy are also known.

It is not too much of an exaggeration to state that future direction of industry, transportation and agriculture depends on how wisely the countries of this region utilize commercial, traditional and renewable sources of energy. The countries of the Pacific Basin, poised at the start of an era of unprecedented economic growth, are particularly well placed not to repeat the old mistakes. Experience has shown that the ecosystems of the island states are especially vulnerable. With land at a premium and local fisheries dependent on delicate coral reefs, there is no room for costly errors. Destroy a coral reef and it is destroyed for good.

Many countries in this most diverse of all the regions of the world, stand on the brink of a period of unprecedented industrial expansion. The momentum built up in the 1970s has been maintained. At the top of the league are countries like the Republic of Korea whose economy grew by 58% during the 1970s. Countries like Singapore, Indonesia, the Philippines, Malaysia and Thailand are not far behind. These industrializing countries, with their new plant and equipment, are uniquely well placed to avoid the mistakes of the older industrialized countries.

A mix of incentives, regulations and technical options could ensure that the countries of this region avoid pollution problems such as acid rain – the curse of Europe and North America. The warning signs are there. In some Indian cities, for example, air quality is beneath the minimum health standards set by WHO. Even the beautiful Taj Mahal is being threatened by corrosive air pollution. The older industrialized nations have been forced to take the costly option of react-and-cure. For the most part, Asian and Pacific nations are in the enviable position to opt for anticipatory and preventive measures.

One of the important determinants of the success or failure of the development process is the role of energy. The main factor is the share of energy in total product cost and that, of course, depends on the price of energy. How this affects different industries needs to be known before formulating detailed energy policies and estimating the potential for improved energy efficiency.

Energy conservation and energy efficiency

Potential savings with existing plants have been estimated to be of the order of 5–10% for the aluminium, copper and steel industries, 10–20% for petroleum refining and 20–25% for pulp and paper manufacturing. For energy-intensive industries, these levels of savings constitute important economic benefits and, at the national level, can become significant in view of the large share of total energy consumed by the manufacturing sector.

The requirement then is for the newly industrializing nations of the Asia/Pacific region to regard conservation as a hidden fuel. Energy conservation measures bring in their train great environmental benefits, principally lower emissions of gaseous pollutants and particulate matters. In the last analysis, no price tag can be put on clean air.

According to a study carried out for the UN New and Renewable Energy Conference, the current global energy consumption of some 9 billion kW could be cut by as much as 80% without putting a break on GNP growth. This is in part borne out by the experience of FR Germany when, between 1973 and 1980, GNP grew by 20% but energy consumption rose by less than 3%.

On average, the per capita energy consumption in developed countries is 100 times that of developing countries. The danger for the Asia/Pacific region is that, if the newly industrializing nations follow the example of Europe and North America, the disparities in energy consumption will increase, and the environmental damage will increase in proportion.

Energy use in industry

The greatest scope for energy conservation is in industries such as iron and steel, cement, pulp and paper. In these traditional energy-intensive sectors the energy bill can account for up to a half of total production costs. The options for lowering those costs are many and varied. The principal components include:

- encouraging the improvement of processes and the utilization of renewable and non-conventional energy resources;
- increasing education and training in energy management;
- promoting good housekeeping and engineering practices.

The most readily accessible form of energy conservation is the 'cogeneration' process of heat and electric power in an integrated energy system. China has installed 222 cogeneration turbo-generators which save the country 3 million tons of oil each year. In India the potential for cogeneration for existing manufacturing industries has been estimated at 1500 MW per year. In UNEP we have been immensely encouraged by signs that the industrializing nations are becoming increasingly aware of the need to adopt energy-conserving policies.

Transport sector energy consumption

The governments of the Asia/Pacific region will need to give equal attention to energy consumed by the transport sector which accounts for about 20% of total consumption in the developing countries of the region. Readily available measures, such as improving traffic flows, sensible city planning, maintaining road surfaces and promoting vehicle maintenance in cities such as Bangkok and Kuala Lumpur, have the potential to reduce energy consumption by up to 50%.

These examples well illustrate our observation that environment and development are mutually reinforcing. Energy transitions along these lines will enhance development and economic growth, while making the cities cleaner and healthier human environments for both its rich and poor inhabitants. The same applies to energy usage in the countryside where nations are confronted with a different set of problems.

Agricultural energy use

In the Asia/Pacific region the population is forecast to increase from the present level of 2.6 billion to nearly 3.5 billion by the end of the century. In most developing countries, more food can be produced to feed the growing population by doubling or tripling cropping and by increasing the yield of each crop on existing cultivated land. However, this will require more irrigation water, fertilizers and pesticides, as well as increased use of farm machinery for ploughing, sowing and harvesting. The logistics for storing, transporting and distributing agricultural produce will also need to be improved. All these imply increased energy usage.

Agriculture, next to households, accounts for the highest consumption of energy in rural areas. While the intensity of use of farm equipment will increase in developing countries, it will still be a small element of about 7% of total commercial energy consumption in agriculture. Fertilizers and pesticides would account for more than 90% and irrigation 2%.

If Asian countries are to have any reasonable expectation of feeding an extra billion people properly, there will have to be a dramatic improvement in agricultural energy use. At present, far less efficient use is made of energy in the developing countries than in the developed countries. The overall efficiency of energy use in India has been computed to be about 6%, while in the industrialized countries it is 5–6 times higher. Improving the efficiency of energy utilization in agriculture will have significant benefits for the developing countries.

In formulating agricultural development policies, governments should provide incentives and technical assistance to promote efficiencies in total energy use. The oil industry, the major supplier of commercial energy as well as a main source for agro-chemicals, also has an important role of providing information, advice and technical services on the proper use of products to their customers.

It would pay governments too to put a major effort into promoting the cultivation of nitrogen-fixing plants. Intercropping with leguminous species, together with greater use of natural fertilizers such as Asian blue-green algae, could greatly lessen dependence on fertilizers.

Fuelwood shortages – the poor man's energy crisis

I turn now to the most critical energy issue confronting the developing nations in the Asia/Pacific – the so-called 'poor man's energy crisis'. A recent World Bank study finds that 75% of the population in Asia and Africa use traditional fuels – fuelwood, charcoal, agricultural residues and animal wastes – for their domestic energy needs. The standing wood volume in Asia, at 15 cubic metres per person, is the lowest in the world. Already vast areas of India and most of Bangladesh have exhausted their fuelwood supplies. Meeting those energy needs is the most critical aspect of Asia's environment and energy conundrum.

In the early 1970s there was great confidence that renewable energy technologies – biogas, windmills, solar power – would provide the answer. Studies prepared for the 1982 UN Conference put the potential for new and renewable energy at a staggering 278 billion kW. But progress in tapping this virtually unlimited potential has been disappointingly slow. A kaleidoscope of social, financial, logistical and technological barriers have stood in the way of widespread adoption of these renewable energies. According to the Stockholm-based Beijer Institute, the realistic projection for new and renewable sources of energy seems to be that it will provide less than 2% of global energy needs at the turn of the century.

Countries in the region must therefore come to terms with the fact that traditional energy needs, now and for the foreseeable future, will have to be satisfied by traditional means. This is but one example among many which illustrates how little the rural poor have benefited from the development process.

The 'top to bottom approach', has, on the whole, been a failure. Take woodstoves. More efficient woodstoves have been seen as a way to relieve pressure on tree cover. Major programmes have been launched in India, Indonesia, Sri Lanka and Nepal. But the stoves which worked so well in the laboratories have failed to catch on in the field. They failed to capture the enthusiasm of the local people who did not have the materials at hand to make their own versions and who valued the 'lost' heat from open fires for ambient warmth and the smoke for keeping thatch roofs free from parasite infestation. No matter that open fires use much more fuel, no matter that the smoke is a threat to health, they still prefer the traditional methods. The experience with stoves is a writ for energy development as a whole. Unless you involve local communities, find out their needs, cultural and economic, no rural energy strategy can be successful.

The UN Energy Conference called for a five-fold increase in treeplanting to meet the energy needs of the 2 billion people estimated by FAO still to be reliant on traditional energy by the turn of the century. Nothing much has been achieved in this direction. In 1984, most energy studies in Asia and other developing regions are still placing greatest emphasis on the village woodlot, cared and tended by the villagers themselves.

But we do have enough success stories, from northern India and from parts of PR China, to show that renewable energy apart from biomass does have a future. The need is to frame realistic policies that will enable the poorer countries of the region to meet traditional energy needs, while also providing scope for the widespread introduction of intermediate technologies. In the rural areas of the Asia/Pacific region the success of such a strategy will help decide the environmental fate of the region.

The need for adjustment

In summary, this region faces the need to embark on a period of thorough-going adjustment. The adjustment process entails a range of actions which must include:

1. undertaking a vastly expanded and more diversified programme of investments to develop properly and carefully indigenous energy resources such as natural gas, coal and peat, taking all the environmental aspects of such development fully into account;
2. promoting research and development for the socially acceptable application of renewable energy;
3. reorienting industrial, agricultural and transport development strategies to take into account higher energy costs;
4. increasing the efficiency of energy use, through rational pricing and changing manufacturing processes;
5. lastly and most important, promoting energy conservation measures.

The oil price increases have helped underscore the interrelatedness of energy resources, development and environment. It has forced resource consumption patterns and development objectives and patterns to be re-examined. More prudent conservation and management of resources have directly contributed towards improving environmental quality.

The experience of the past decade has confirmed that considerable potential exists for reducing the energy intensity of economic activity. Countries in Asia and the Pacific will be able to draw upon and benefit from this experience. There are opportunities to choose the most appropriate strategies to balance energy consumption, development goals and environmental quality.

Increasing standards of living, while at the same time improving the quality of life, is a realizable objective for all the countries of the Asia/Pacific region. What is needed is a solemn resolve to work effectively towards achieving this goal, both separately and as a region.

22

The premises for building a sustainable society

Address to The World Commission on Environment and Development

Geneva, Switzerland, October 1984

In the period since the Stockholm Conference in 1972 our ideas about the relationship between environment and development have undergone a fundamental change. Far from being in conflict with economic growth, the environment is now seen as its friend. No one now challenges the concept that we should use the world's natural resources sustainably. Information, ideas, and theories sufficient to fill a large library have emerged, yet the fact remains that those ideas have too seldom been applied.

The result is that, in 1984, the four main biological foundations of the global economy – croplands, grasslands, forests and fisheries – are under greater strain than ever before. We have the know-how and most of the means to build a sustainable global society, and yet we appear, in many areas, to be going backwards. This is all very perplexing and frustrating. It is as though we confront the Minoan maze but lack the golden thread to guide us to the centre.

This has been a prevalent feeling at UNEP's Governing Council and in its Secretariat. How far have we advanced in achieving a common understanding of environmental issues? Where are we heading? And what are our ultimate goals? In an endeavour to provide the answers the decision was taken by the Governing Council to prepare an Environmental Perspective to the year 2000 and Beyond. For this purpose, the Council established its own Intergovernmental Preparatory Committee. The Perspective document would be presented to the General Assembly for adoption. A Special Commission, comprised of eminent international figures, would assist the Governing Council, in addition to its work in addressing the larger issue of environment in the global *problématique*. The General Assembly endorsed the Council's recommendations, generous funding from various governments was pledged and the Chairman and Vice-Chairman were appointed by the Secretary General of the United Nations.

In UNEP we too are committed to establishing a fruitful, harmonious relationship with the Commission. The General Assembly stipulated that close ties should exist between the Commission and the Intergovernmental Inter-Sessional Preparatory Committee, set up by the Governing Council to prepare the Perspective Document.

Premises for building a sustainable society

Though the idea for a Commission evolved within UNEP, and though it will fall to our Governing Council to transmit your report to the General Assembly, it is not part of our function to say how the Commission should tackle its job. But I thought it might be helpful if I devoted the major part of my statement to outlining what we see as the premises for building a sustainable society.

UNEP is convinced that progress in tackling environmental despoilation depends on a more thorough understanding of its linkages, direct and indirect, with the wide aspects of economic and social development. The environmental affects and is, in turn, affected by the forces which have created and still govern the global *problématique*.

But when, from an environmental standpoint, we attempt to identify those forces, the threshold into the maze is crossed. The forces include:

- the mismanagement of our natural resource base;
- poverty and its impact on human beings and their environment;
- overconsumption and the impact of the demands of the privileged on the environment;
- population growth, especially in those areas where the increase in human numbers is undermining the carrying capacity of the biosphere;
- the squandering of human and natural resources on the development and manufacture of armaments;
- the widespread lack of understanding on the part of decision makers and the public at large of the implications of environmental destruction for social, economic and political stability.

What, then, are the premises for possible solutions? From what base will we move the ideas for sustainable development into action?

The first and most important premise is the generally agreed perception that economic development and environmental quality are interdependent and, in the long term, mutually reinforcing. The rational management of the world's threatened natural resource base forestalls a loss in environmental quality and enhances sustainable economic growth. For example, recycling of waste materials serves environmental and economic objectives simultaneously.

A second premise is the need to view environmental problems as a system. A coherent set of solutions are required to ensure they each have a positive impact on the other. For example a hydroelectric dam may be built to generate electricity and provide water for irrigated agriculture, but both objectives – more energy and more food – will be frustrated in the long term unless steps to conserve watersheds and to avoid eutrophication and salinization are taken during the planning and construction stages.

A third premise is flexibility and anticipation. If we start from the premise that today's problems will also be tomorrow's, then we needlessly close future options. We need to keep a weather eye on problems as they evolve. We need to devise a long-term and flexible response that will help in overcoming the inevitable time lag between agreeing solutions and applying them. Fundamental to this objective is a system for monitoring and anticipating environmental trends. Satellite monitoring and other techniques have vastly improved our capacity for observing environmental changes. Within limited manpower and human resources, UNEP's Global Environmental Monitoring System has made important progress, providing decision makers with the hard data they need for forward planning. If we are to anticipate future changes successfully, a great deal more attention will need to be given to environmental monitoring.

A fourth premise is that actions should be sufficiently dynamic to have a meaningful and positive impact on the whole system of environmental problems. The implication is that processes in the *problématique* which are, by their very nature, capable of having such an impact and of being directed, must be identified and harnessed.

The large-scale and dynamic process in the *problématique* which may be harnessed to resolving environmental problems, and which, maladdressed, causes them, is development. The idea, thus, is that development should be harnessed to solve the problem of the global *problématique*, and should do so in a stable manner over the long term. The capsule phrase given to this idea in UNEP is sustainable development.

The fifth premise must be a much greater awareness among the public and decision makers, in particular, of the environmental dimension to the global *problématique*. Without public pressure for action, little that is positive will be achieved. Unless people, in both developed and developing countries, understand more fully what is at stake and how their future well-being is threatened, our objectives will not be realized. Implementing sustainable development strategies will demand trade-offs and, sometimes, short-term sacrifice. It is only through greater public understanding that a political climate will be created that will allow decision makers to take the required actions.

I think it is fair to say that increasing numbers of the public are aware that the pressures on the environment are mounting. The loss of tropical forests, species extinctions, the encroaching deserts, acid rain, nuclear waste disposal, dangerous pesticides, chemical dumping, soil erosion and the human settlements crisis are all environmental issues that have aroused a good deal of public concern. But I think it is also fair to say that, beyond the environmental constituency, there is little or no understanding that these environmental problems are merely symptoms of more deep-seated, intractible forces, and that these are generated by insufficient or poorly executed development.

There is even less understanding that we are all in this together. The consequences of environmental destruction pay no heed to social, political and economic barriers. A more thorough understanding of environmental interdependence between nations is an essential precondition for building a sustainable society.

Until recently we in the environment movement have been very good at pointing out what is going wrong with the world. Our case for sustainable development has rested squarely on the negative. But there are signs that we are moving into a more pragmatic, more clear-eyed phase. During the debate on desertification at our most recent Governing Council, it was notable how government after government called upon UNEP to point to instances where schemes to combat the advancing desert had been successful. They wanted some encouragement; they wanted, in other words, to see sustainable development in action.

Tasks and problems facing the Commission

The major tasks facing the Commission are clear: to conceptualize sustainable development in operatonal terms and to develop long-term strategies for its achievement. These strategies must appeal and have meaning to people who may have little or no knowledge for the environment. Here I am talking of the planners, economists, businessmen, politicians, lawyers, accountants, soldiers and engineers whose decisions decide the course of development, and thereby the fate of our human environment. With such a course embarked upon, the golden thread will surely begin to unwind.

To mean anything these strategies must bring hope to millions who live on or near the margins of existence: 2 billion people who must chop firewood to meet their meagre energy needs; the rural and urban majorities in developing nations who lack access to safe water and sanitation; the 850 million who live in the shadow of creeping deserts; and the as yet uncounted millions of drought victims who have eaten their seed grain and slaughtered their breeding livestock. They have been forced to discount tomorrow in the struggle to survive today. Their lifestyle is one of self-immolation. If we fail to answer their needs, if we fail to point out ways and means they can become beneficiaries of development, then this exercise will have been a futile waste of time.

I believe you would agree that, unless the underprivileged are made to feel that they have a full role to play in development, no strategy for sustainable development can hope to be successful. This means that the Commission will need to confront some thorny issues; and the strategies suggested will not always be palatable.

Foremost will be the issue of improving land tenure. When a farmer feels insecure, the tendency is to 'mine' the land until it is no longer productive, and then move on elswhere to repeat the process. As a recent World Bank Report stated: 'There is vast potential in improving inequitable land tenure patterns which promote poverty, environmental degradation and inefficient natural resource use'. Surely this is an absolutely crucial factor. For unless the underprivileged are given security of tenure, little headway will be made in improving environmental management in the rural areas. Much the same applies to dealing with the developing crisis in the world's urban areas.

The daunting prospect facing the global community is a population stabilizing at around 10 billion people midway through the next century. We must expect that, in the short term, the numbers of the absolute poor will increase, making more difficult the aim of achieving a balance between environment and development.

This need not be a cause for despair. The increase in population is essentially a result of our success in conquering many diseases and in providing a greater quantity and variety of food. With the notable exception of Africa, per capita increases in food output have kept pace with the increase in human numbers. The challenge is two-fold: to identify the causes of environmental destruction and the elements of success. We need to weed out the former and build on the latter.

A problem the Commission is certain to encounter is that, in general, neoclassical economics and environmental management are difficult to merge, mainly because environmental values are difficult to quantify. Environmental consequences, such as air quality, soil fertility and so on, have important implications for human health and agricultural productivity and, although these may be important economic factors, they are not amenable to exact measurement. How, for example, do you put a price tag on keeping a river unpolluted, or on preserving an endangered species that may turn out to be of major economic value? What is the economic justification for conserving slow-reproducing renewable natural resources such as whales, redwoods or tropical hardwoods? The classic economic approach would be to destroy them completely and invest the profits in a more lucrative enterprise.

However, when we see how marine pollution and destruction of fish breeding grounds is undermining marine-based economies, when we see how deforestation is affecting local climatic patterns, and when we see soil erosion undermining agricultural productivity, we see also the literal bankruptcy of the traditional approach which measures successful development purely in high GNP growth rates. Such an approach creates powerful incentives for countries to 'liquidate' their natural resource base as rapidly as possible.

What needs to be achieved is an accommodation in economic decision making of the long-term social, aesthetic, spiritual and, of course, economic consequences of the irreversible liquidation of natural capital. It all comes down to keeping the options open for ourselves and future generations. In short, the need is to define the specific principles of sustainable development and couple them with current economic theories and practices.

Understanding global systems

I do not underestimate the difficult burden the Chairman and members of the Commission have assumed, a burden made all the greater by the complexity and the fog of uncertainty still surrounding the relationship between people and the systems of the world. Though we have learnt a great deal, it has served merely to show how much we still have to learn about the relationships and linkages between people, resources, environment and development. Take the term 'carrying capacity', so much in fashion these days. It is not a fixed concept, but metamorphoses as human numbers increase, as climate shifts, as ecosystems evolve and as development patterns change.

Not in the next few generations will the global systems be so fully understood that they may be treated mechanically, and, in prescribing solutions, a sense of humility is required. Our approaches should be tempered by awe of our natural and social inheritances. A spirit of not destroying what we are yet to comprehend and a spirit of symbiotic partnership with all that matters on our planet should guide our actions. This too is an aspect of 'sustainable development'. Strategies for sustainable development must look to people, motivate them and give them the tools to achieve their aspirations. To deal with the issues of environment thus becomes a matter which transcends environmentalists, who must look to the conviction they carry with humanity as a whole as they pursue their economic, social, military or reproductive activities. Conceptualization must start from two basic perceptions:

1. Life processes are interconnected in intricate ways.
2. Air, water, land and life constitute an interlocking system; elements vital to all life move in cycles between the rocks, water, air and living matter. Harsh experience has shown that, beyond certain limits, these cycles cannot be disturbed without causing irrevocable damage. A recognition of these limits is at the heart of environmental management. We must use nature but we must maintain natural systems sufficiently intact to sustain genetic richness and to maintain the supportive cycles of the biosphere.

It is not only that the life processes are interconnected, but that the seamless web spun by today's trade, communications and finances has also submerged the interconnections in the world's economic and political systems. These interconnections are increasingly placed in jeopardy by the conflicting demand for resources and the growing divergence of need, interest and power among countries at different stages of development. It is necessary, in the interest of all, to seek adjustment of these differences so that sustainable development of the poorer regions of the world receives first priority. Otherwise, population is less likely to be stabilized and political tensions and pressures on resources will inevitably increase.

Formulating strategies for sustainable development

In formulating appropriate strategies for sustainable development, it will be necessary:
1. to evaluate the realistic options available to countries and groups with different sociopolitical systems and at different stages of economic efficiency and technological advancement;
2. to disseminate information on environmental and conservation problems so that decision makers at all levels, and in different sectors, move from react-and-cure to anticipatory and preventive policies;
3. to make explicit the linkages that exist between development and low-waste and non-waste technologies, recycling, industrial and energy planning, and so forth.
4. to achieve a number of critical transitions: an energy transition to an era in which energy is produced and used at high-efficiency without aggravating other global problems; a demographic transition to a stable world population of around 10 billion; a resource transition to reliance on nature's 'income' and not depletion of its 'capital'.

The Commission was seen as an essential means to achieve the task because governments in the Governing Council of UNEP and in the General Assembly felt that the eminence of the participants, working in substantive independence, and mobilizing great resources of experience in all the areas of the global *problématique*, would not only develop the necessary approaches but also carry conviction with the world community. This would greatly assist governments in their own tasks, intergovernmentally, of reaching the required agreements and implementing them.

UNEP initiatives

The Commission may well ask – how far has UNEP itself progressed in this task? Since the Stockholm Conference and the creation of UNEP, many initiatives have been taken, and I would like to describe the principal ones as stages or benchmarks in a process, and share with you some conclusions.
   1.   At the Stockholm Conference itself the concept of poverty as a major pollutant was made explicit for the first time. It was widely accepted that the pressures on life-support systems generated by poverty are as significant as the pollution created by industry, technology and overconsumption by the affluent: both lead to the rapid depletion of basic natural resources. The conclusion reached in terms of sustainable development was the need to internalize environmental constraints in development.
   2.   At the joint UNCTAD/UNEP Symposium on Patterns of Resource Use, Environment and Development Stategies held at Cocoyoc in Mexico in October 1974, the internal and international linkages of patterns of consumption in developed and developing countries were considered and the concept of 'ecodevelopment' was developed. According to the Cocoyoc Declaration, the road ahead lies 'through a careful and dispassionate assessment of the outer limits, through co-operative search for ways to achieve the inner limits of fundamental human rights,

through the building of social structures to express those rights, and through all the patient work of devising techniques and styles of development which enhance and preserve our planetary inheritance'.

3.   I pointed out to the World Food Conference in the same year that any strategy to increase food production on a sustained basis should explicitly take account of the complementarity of environment and development. The objective must be to maximize food production without destroying the ecological basis for sustainable production. I noted that strategies to solve the world food problem must be developed in full knowledge of the web of interdependence that exists between food production and the other major problems facing mankind.

4.   A series of seminars were organized by UNEP in cooperation with the Regional Economic Commissions during 1979 and 1980. The regional seminars clearly showed that to reach meaningful conclusions on alternative development patterns and lifestyles, attention must focus on a broad range of interrelated issues including: technological options and resource profiles; patterns of rural development, including, particularly, food systems and management of renewable resources; patterns of urban development; and, above all, questions relating to institutional aspects of management and participatory planning. These conclusions were considered, in turn, by the executive secretaries of the regional economic commissions and experts who met in Nairobi in March 1980. They made recommendations to the Preparatory Committee for the New International Development Strategy. These Recommendations considered that 'humankind is a part of the biophysical world, acts upon it and is affected by its reactions. The biophysical world is the life-support system of society and provides space, a flow of materials and energy, and a medium for the reabsorption of wastes. These functions of the environment, adequately understood and wisely managed, constitute a basis for the achievement of the goals of development. It is essential, therefore, that the interaction between patterns of development and the environment be fully and explicitly taken into account in the New International Development Strategy for the 1980s'.

These and other recommendations from UNEP were taken into account in the framing of the New International Development Strategy (NIDS) for the Third UN Development Decade, adopted by the General Assembly in December 1980. The present NIDS stresses the need for a development process which is environmentally sustainable over the long run and protects the ecological balance.

5.   In 1980 the World Conservation Strategy, produced by IUCN in partnership with UNEP and WWF, was published. It set out clearly the objectives for living resource conservation: first, to maintain essential ecological processes and life support systems; second, to preserve genetic diversity; and third, to ensure the sustainable utilization of species and ecosystems.

6.   A decade ago, following upon the first World Population Conference in Bucharest, the General Assembly called for a programme of studies on the interrelationships between population, resources, environment and development. But it quickly became clear that the conceptual base was lacking, and UNEP turned its attention to building such a base. A symposium on interrelationships was held at Stockholm in August 1979. The Symposium was followed by two UNEP-convened sessions of a high-level expert group on the interrelationships, in 1980 and 1981. The group developed the required conceptual approach and its major recommendations, subsequently endorsed by the Governing Council, ECOSOC and by the General Assembly itself, were as follows:

● The exercise should be on the interrelationships between people, resources, environment and development. While the demographic variable is an important dimension, the issue of making people effective participants in and beneficiaries of the development process is central.

- A systemic approach should be tried in a limited number of geographically or otherwise distinct areas identified by the experts.
- A system-wide programme of work should be established, together with a voluntary fund to support it.

At present, the UN system is working on two case studies, one on the deforestation of the Himalayan Foothills, and one on carrying capacity in Kenya.

In a major contribution to the 2nd UN Population Conference, UNEP outlined its position on the role of population increase as a causitive factor in resource overexploitation. This statement, and also the section of the 1985 State of the Environment Report, consider the population and environment issue in detail.

7.   A relevant initiative taken by UNEP is our Programme on Peace and Security and the Environment. We sponsored the Galtung Report which you may wish to examine. I addressed the Special Session of the General Assembly on Disarmament and drew attention to the fact that the traditional military concept of security was becoming increasingly obsolete. It must be broadened to include such threats as the depletion of natural resources and a deterioration of the living environment. A commitment to the concern of environmental betterment must be seen as a long-term commitment to national and global security. The long-term cumulative effects of even a limited nuclear exchange on the environment, the so-called 'nuclear winter', has recently demonstrated the magnitude of such disruptions and their irreversible nature.

8.   UNEP has undertaken the development of analytical tools to make clear and more effective the integration of environmental considerations in development projects, programmes and planning. Our current programme of work includes the preparation of cost-effective and simplified formats for environmental impact assessment, the application of cost–benefit analysis to environmental protection measures, the adaptation of integrated physical, socioeconomic and environmental planning to the requirements of the developing countries, and other tools. One area to which particular attention is being paid is the linking of environmental and resource data to traditional economic data (the production of goods and services) through the integration of environmental statistics in the System of National Accounts. In this way, 'satellite' accounts could be constructed which will show environmental and natural resource losses and gains as a supplement to financial indicators. There will be a joint World Bank/UNEP meeting in November this year to develop guidelines on environmental accounting for use by developing countries.

9.   At the Session of a Special Character (SSC) of the UNEP Governing Council in 1982, the Nairobi Declaration was adopted. In that Declaration both the poverty of the South and the wasteful consumption patterns of the North were described as threats to the environment, since both could lead people to overexploit their natural resource base and the regenerative capacities of nature. Emphasis was laid on the interconnection and interrelationships between environmental problems and concerns. The SSC urged governments, *inter alia*, to establish or strengthen their national mechanisms for the integration of environmental considerations into development planning.

10.   The 1984 State of the Environment Report considered the role of the environment in the dialogue between and among developed and developing nations.

It shows how environmental concerns are central to the political, economic and social issues on which the dialogue concentrates. It puts the case for environmental interdependence by arguing that economic problems and widespread poverty force people to misuse resources and degrade their environments. This, in turn, makes economic growth and reforms harder to achieve.

Conclusion

I believe that UNEP's initiatives in this field could provide you with some building blocks. I would like to pledge UNEP's fullest possible assistance with the Commission's endeavours. We have the utmost confidence that the report of the Commission will provide the international community with a new vision and a new sense of purpose as we work to build a sustainable global society.

23

Partnership for conservation

Statement to the Sixteenth General Assembly of the International Union for the Conservation of Nature and Natural Resources

Madrid, Spain, November 1984

UNEP welcomes 'Partnership for Conservation' as the theme for this Sixteenth General Assembly. The close partnership between UNEP, IUCN and WWF has been a source of satisfaction to all of us. And we can fairly claim a good deal of the credit for broadening the base of support for conservation-based development.

Our World Conservation Strategy has spawned 33 national strategies, the Charter for Nature has been adopted by the UN General Assembly, there now exist 114 international environment treaties. At the time of Stockholm there were less than a dozen environment and natural resources machineries in governments; there are now more than 100. All major multinational development assistance institutions are now formally committed to environmentally-sound development. The list of achievements is long, and is getting longer.

Putting ideas into practice

Although we can take pride in our achievements, they must be placed in context. No one now challenges the concept that we should use the world's natural resources sustainably. Information, ideas and theories sufficient to fill a large library have emerged, yet the fact remains that those ideas have too seldom been applied. The consequence is that, in 1984, the four main biological foundations of the global economy – croplands, grasslands, forests and fisheries – are subjected to greater strain than ever before. We have the know-how and most of the means to build a sustainable society, and yet we appear, in many areas, to be going backwards.

It is with this in mind that UNEP wholeheartedly welcomes the general thrust of the proposed conservation programme for 1985–87, so ably worked out by the Director General of the IUCN and his staff. We also welcome the proposed World Conservation Plan, intended to identify activities required to make the World Conservation Strategy work on the ground. In particular, we support its objective of determining what role each organization can play in furthering the implementation of the Strategy.

Linkages between environment and development

In working out the Plan, I am sure all partners will take fully into account our conviction in UNEP that progress in tackling environmental despoilation depends on a more thorough understanding of its linkages, direct and indirect, with the wider aspects of economic and social development. The environment affects and is, in turn, affected by the forces which have created and which still govern the global *problématique*. Those forces are:

- the management of our natural resource base;
- poverty and its impact on human beings and their environment;
- population growth, especially in those areas where the increase in human numbers is undermining the carrying capacity of the biosphere;
- overconsumption and the impact of the demands of the privileged on the environment;
- the squandering of human and natural resources on the development and manufacture of armaments;
- the limited understanding on the part of decision makers and the public at large of the implications of environmental destruction for social, economic and political stability.

Cooperation and collaboration

UNEP is well satisfied with the results of its close collaboration with IUCN, drawing as it does on the scientific community all over the globe. Although the resources available to our Environment Fund have shrunk in real terms by 50% since 1978, we are continuing to support generously activities within IUCN's technical and scientific remit. I pledge UNEP's firm commitment to continue support to this useful cooperation in the future, within the framework of our approved programme and what you will approve at this session as the programme for 1985–87.

Close partnerships, such as those that exist between IUCN and UNEP, are bound to change and evolve. Under the new arrangements concerning the Secretariat of the Convention on International Trade in Endangered Species, CITES, we will continue to maintain the closest collaboration with IUCN. Specifically, we will be looking to IUCN to continue its scientific support which is so necessary to the effective functioning of the Convention.

Support for environmental issues

To date, few global economic prescriptions for overcoming the economic depression – which vary from a complete overhaul of the global economy to leaving it alone to settle – have paid much heed to the biosphere. We have yet to convince a majority of governments effectively that the environment is essential to economic and social development. They remain locked into existing patterns of development in which environmental considerations are conceived to play only a peripheral role. This is a course which, sooner rather than later, forces them to take costly remedial measures. We must therefore deepen and broaden our partnerships to make governments, industry and all other societal groups take more account of the resources of the biosphere upon which nations' economic and social welfare is so completely dependent.

We have on our side one vital resource – mushrooming support among the general public. Time and time again we have seen public pressure forcing the hand of government and industry. As numerous opinion polls testify, the public is profoundly concerned about acid rain, the disappearing tropical forests, and pollution of our coasts and rivers. Typical is a recent poll in FR Germany which found that 98% of the people canvassed believe that improving environmental quality should be their government's most important priority. I believe we in the environment movement should be doing much more to utilize public support in our dealings with decisionmakers. UNEP will take particular note of any recommendations this Assembly makes in this regard.

I would like to pay tribute to the NGOs represented here for their role in arousing public interest and concern, and also to pledge UNEP's full support to the welcome initiative of WWF's Director General to place more emphasis on making as many people as possible aware that wise management of natural resources is a top priority for meeting basic human needs.

Drought and famine in Africa.

The current crisis in Africa brought on by the drought has exposed the true extent of the environmental crisis. For more than two years now, the UN and relief agencies have been warning that famine would strike when the next drought came. Since 1977 UNEP has been warning that, if desertification was left unchecked, Africa would experience a disaster worse even than the terrible Sahelian drought in the early 1970s. Yet it took television picures of the famished, and the widespread public concern they generated, to make governments respond to anything like the scale required.

It is estimated that during the current drought, two and a half million hectares of grazing and cropland have been destroyed. How much of that is irreversible we do not yet know. We do know, however, it is but an acceleration of the process of land degradation that has taken place even in years of good rainfall. We pray that the current drought will soon be over, and from some regions there are signs that it may be. But, inevitably, another drought will occur, and this time nations have to be prepared if they want to avoid a crisis even more severe than the current one.

The danger is that a return to normal rainfall will divert attention from this hidden environmental crisis. Hidden because its side effects – displacement of rural populations, per capita reductions in food output, political and social unrest – are seldom traced to environmental impoverishment. Now, when everyone is asking how do we avoid a future crisis, we environmentalists should be saying 'here are our ideas, plans and strategies, apply them now because you may not get another chance'.

But any ideas, plans or strategies are worthless unless they squarely address the needs of the poor. The malnourished, the diseased and the dispossessed cannot be expected to plan ahead. When the millions who live in the path of encroaching deserts, the millions who see their land being washed and blown away, and the millions who live daily with the fear of landslides and drought are given a stake in the future, then, and only then, will we be on the road to a sustainable society.

Acting in partnership

While we might be encouraged that groups and organizations whose brief is not environmental are coming to embrace our cause, we must also add our weight and influence to theirs. This is the new direction our partnership should take. Family planning, drought and disaster relief agencies, youth associations, labour unions, consumer protection organizations, public health and women's groups can benefit from reinforcement by the environment community.

Our voice can also have great effect on the disarmament movement, in the UN and elsewhere. We must make our full contribution to the forces of reason and sanity which refuse to accept the inevitability of the arms race. For while we continue to tolerate the deflection of more than a third of the world's research and development expenditure into preparing for war, we will never make peace with ourselves or with the earth.

In one of her last public statements in support of conservation, India's Prime Minister and Minister of Environment, Mrs Indira Gandhi, made an appeal for man to retain a communion with the earth. She said: 'We must enable the earth to renew itself. We must aim to improve the material, intellectual and spiritual circumstances of peoples. And we must nurture the values which enhance human possibilities. Our ancients believed in the unity of all living things, and even of life and non-life. We must rediscover the sense of identity with and responsibility for fellow humans, other species and future generations'. I can think of no more appropriate message to send from Madrid to the outside world.

24

Environmental management

Statement to the Opening Session of the World Industry Conference
on Environmental Management

Versailles, France, November 1984

Today one fact of human life is more certain than ever: the stakes involved in finding
solutions to man-made environmental problems are no less than the future and fate
of our planet.

Over the previous decade, industry, governments and the environment community
have come a long way. Only extremists still see ground for confrontation. In 1984 the
burden of proof rests with those who believe economic growth can be generated
without conserving resources – and the silence from that quarter has been
overwhelming. So let us be clear from the outset, we are here not only to create
common ground; we are here not only to ask why, we are here to ask how – how,
with a strong commitment to close cooperation, we can promote a development
process which husbands our shared resources. The fact that high-level industry,
government, parliamentarians, the scientific community, labour and NGO's share
this hall today augurs well for achieving this goal.

The environment is a story of complex interdependencies, just as the ecosystem
itself involves a series of delicately balanced forces. We must keep that balance in
mind, and address squarely the interdependency of development and environmental
protection, which is so great that neither can be separated from the other. In fact,
this conference has been convened because, consciously or unconsciously, we all
recognize that relationship. The need to establish a mutual reliance is no longer in
doubt; what this conference is about is to suggest ways we can strive to make it a
working relationship.

The discussion papers before you try to put together – based on your own
contributions to the preparatory process – where we stand, and to identify some
issues you may wish to consider. There is no prejudgement of what will come of this
World Industry Conference on Environmental Management (WICEM), or of what
its conclusions will be. What comes out of the conference is what you, and only you,
decide upon. The privilege in this conference is ours; we the theorists of environment
and development value the opportunity to talk to the practitioners. If the groups
represented here do not emerge in greater harmony three days from now, then
WICEM will be judged a failure.

There is mounting evidence that excessive demands are being made on the world's
limited resources and on the productive capacity of ecosystems. World industry
relies on developing countries as the main source of many raw materials, but already
the renewable resource base in many developing countries has been seriously
undermined. Many examples exist of governments which undertake environmentally
ill advised projects with several multinational corporations; certainly both sides

share the blame for the ensuing environmental damage, but ultimately both sides must be persuaded that it is not in either of their interests to pursue such shortsighted economic policies.

It is a question of survival in more ways that one: as we destroy basic resources, we run the risk of aggravating already tense political situations. If war over oil is conceivable, then why not war over water? World-wide, 145 river systems are shared by two or more states; already we have seen how pollution and unsustainable withdrawals have created friction, and the pressure will grow. Using current technology, industry will require four times more water by the year 2000 to keep the same number employed as in 1975. Food shortages and famine certainly show up the need for much more and better agricultural development. This will mean escalating demand for water. Aspirations for better standards of living, especially in developing countries, is another mounting claim on water for domestic use.

Since the 1972 Stockholm Conference on the Human Environment, the environment community has produced information, ideas and strategies sufficient to fill a large library – yet the fact remains that those ideas have too seldom been applied. The explanation is undoubtedly that we have been guilty of playing to our own gallery, moving, if you like, too much in our own bubble. I would venture to suggest that so many representatives of industry have come to Versailles because they recognize they have been doing much the same thing.

We know that when we want to work together for a desired goal, social groups will find common ground. Each of the social partners participating in WICEM has brought with them attitudes, perceptions and approaches to our common problem. We can all benefit and learn from each other.

The priority of this Conference's deliberations should be to concentrate on those issues that unite government, industry and other sectors of society in the cause of environmental management. However, we should have a careful look at the world's military industrial complexes which absorb one third of the world's research and development resources. We, in the environment movement, are committed to adding our weight to those who argue that arms investment does not provide nations with meaningful security but that investment in resource conservation will.

The context for the Conference is the certainty that ecology and economics are mutually reinforcing. Its challenge is to identify ways in which cost-effectiveness, public policy and environmental necessity can all be fused in operational and management terms. WICEM can become the foundation of a closer, more fruitful relationship between all partners concerned.

At the industrial level, there is a natural reluctance to invest in 'fixed costs' rather than in increased production. But industry must be alerted to the long-term consequences of this reluctance, which will cut into their profits just as surely as it will harm the environment. The challenge is to start people thinking of long-term costs and benefits using environmental criteria as important factors in the equation.

In some business circles there is a belief that the regulatory burden of environmental measures has grown so enormously in recent years that they impose 'unreasonable costs' for 'questionable benefits'. Environmentalists counter that, although the initial cost of regulation is high, it is not as high as claimed, and in any event it is offset by short- and long-term benefits. Already private environmental management firms are creating jobs, generating profits and serving the cause of environmental management out of self-interest – the best motivation.

A mainstay of sound environmental management is that anticipation and prevention of environmental difficulties are far preferable to reaction and cure. This is cost–benefit analysis of a sort that businesses consider every day in the course of their operations. To remain viable and profitable, industry must forsee change, not react to it. Here, we in the environment movement have a massive contribution to make by providing industry with early warnings on environmental trends – for example, on rates and likely consequences of marine pollution, CO_2 build-up, desertification and tropical forest destruction. The enrionment movement may also use its influence with governments and parliamentarians to forge laws and regulations better suited to various societal groups, including industry.

Because it is not intrinsically a political problem, management of the environment is an apt point of departure for improving the dialogue between East and West, North and South. It is not too much to hope that successful cooperation on environmental matters could lead to an improvement in the climate for negotiations in other areas of the global *problématique*. WICEM provides us with an opportunity to make great strides in this direction.

Within this wider context, industry's technical expertise and management skills are vital to the solution of many complex and interrelated environmental problems. By sharing that experience and knowledge, particularly with developing countries, industry can ensure that economic development is realized on the basis of sound environmental management. We should do more to use the revolution in information technologies to make sure that experience and knowledge is made more freely available.

We are also here to see to it that the scientific and economic information from the combined experience of industry, government, and all the groups represented at WICEM, reaches where it is most needed: the national and local levels, right down to the factory lines and farmers' fields. To make this transfer of information a reality we must forge an alliance of cooperation between the groups assembled here, and see the process through to its end. We are not starting from scratch: for nine years UNEP, through its Industry and Environment Office in Paris, has engaged in active consultation at the global and regional levels. But WICEM's real challenge is to see if we can instigate similar kinds of discussions at the national level, where the most meaningful action can be taken.

Yet another challenge faces us here. Most member states of the United Nations have large public sectors and government-run industries, but a greater willingness must be shown on their part to accept incentive systems and industry self-regulation. Industry, on the other hand, may have to rethink the level at which an enterprise is considered to be 'profitable' and the extent to which they are prepared to accept the constraints under which various political systems operate. I am confident that the conference will suggest ways in which such adjustments can be made.

We can use this conference to make the 1980s a decade of action-oriented partnership between governments, the industrial community, and all other sectors of society. Our goal should be to see to it that the cooperation engendered by WICEM does not evaporate in the months and years to come. Our obligation is to the environment and to future generations. We are all aware of the magnitude of the problem; let us ensure that our response is equal to the task.

25

Solving our environmental problems

Statement to the Opening Session of the
Inter-parliamentary Conference on Environment

Nairobi, Kenya, November 1984

In recent years, we have attempted to begin a dialogue that recognizes the interrelatedness of environmental issues and the interdependence of states in dealing with their major environmental concerns. The environment is like a jigsaw puzzle; if a single piece is missing then the picture will not be complete. If, for example, some nations are polluting the seas, it is not just these countries that will suffer. All contiguous nations will have to bear the burden of environmental destruction. Destruction of tropical forests can have detrimental effects on the economies, and perhaps even the climates, of European and North American countries.

Dialogue between states and between various societal groups within countries is crucial to achieving meaningful solutions to our environmental problems. Our most recent endeavour in this respect is the World Industry Conference on Environmental Management concluded just a few days ago in Versailles, France. Its participants came from governments, industry, parliaments, and labour and non-governmental organizations. The Versailles Conference produced a strong consensus among participants that environmental protection and economic growth are compatible. It also recognized the need for environmental planning to reach every level of industrial management. The declaration of the Conference is available to you for your consideration.

We consider the present Conference represents a further, and equally crucial, forum for advancing the dialogue – this time among parliamentarians, and between them and scientists and environmental technicians.

We know we are still very far from the goals set for us by the international community at Stockholm in 1972. However, through dialogues such as this, some remarkable results have already been achieved. First, international mechanisms for consultation and for forging consensus between states have been applied. During the last 12 years, 32 international environmental conventions and protocols have been signed under the auspices of the United Nations. Second, specific control programmes have been drawn up, with most attention given to dealing with regions or systems particularly at risk – watersheds, areas of rapid urbanization, tropical forests, coastal regions and lands in the path of advancing desert. Third, we have promoted a wide range of tools and principles for effective environmental management: environmental impact assessment; environmental cost–benefit and cost-effectiveness analyses; and integrated physical planning and environmental accounting. Fourth, there has been concerted action to ensure that sustainable development is recognized as the only acceptable course for economic growth.

Yet, with all this in place, the biological basis of our economic growth – in fact for our very survival – is being severely and sometimes deliberately eroded. We are deeply concerned that an acre of forest land is disappearing every second, mostly in the South. We are concerned that encroaching deserts are destroying 6 million acres of productive land, and reducing to zero economic productivity 21 million more every year, again almost entirely in the South. In the North the public has joined our deep concern over issues like acid rain. We are all worried about possible climatic changes, with potentially disastrous social and economic consequences. There is certainly a desire at the grass-roots level in all corners of the globe to think of these issues globally and act swiftly at the national level, to use finite resources wisely and to use living resources sustainably.

Parliamentarians are not only legislators. They are the representatives of their constituencies. They can lead their constituents in breaking new ground, in reaching new norms and standards, and in reaching insights and perceptions that become common ground later. Although environment was once considered a vote-loser, this is no longer so. Too often legislators in the West have been surprised by public demands in areas such as emission controls, acid rain and a host of other environmental concerns. Public opinion polls and the rapid growth of citizens' organizations concerned with the environment testify to the fact that environmental concerns are influencing the way people are voting. Simply reacting to particular crises no longer satisfies the public. Elected representatives need not lag behind their constituents in environmental forethought; they are expected to take the lead.

While we are encouraged by the increase in environmental departments and ministries in different countries (from 10 countries at the time of the Stockholm Conference to well over 100 at present), it must be admitted that many of these are understaffed and lack adequate financing. Graver still, very few of them have the necessary political clout to deal effectively with the more established and powerful bureaucracies of financing, industry or agriculture. In the event, decisions are often taken on the basis of short-term economic calculus, disregarding long-term and more fundamental environmental and natural resource consequences. If these environmental bodies are to reflect the actual needs of the country, then your dialogue with their representatives must be deepened, and their support in parliaments must be more visible.

Until now most of UNEP's links have been with the executive branch of government. While that liaison has carried us a long way, the executive has often lacked the powers that could enable it to act on environmental issues. These powers rest with you, the legislative, whose voters, particularly in the North, are demanding sounder environmental management. That demand is growing and expanding into the South, and I would assert that the parliamentarians who are willing to lead their electorate have much to gain. I am also acutely aware that, if those dealing with environment do not have your full support and cooperation, nothing much can be achieved. In a number of countries, framework laws, sometimes prepared with international assistance, have been stalled for many years in the national assembly, or have not been complemented by existing laws. A number of international conventions on environment are still awaiting ratification by parliaments, and 13 conventions or protocols prepared and approved under the aegis of UNEP have not yet entered into force because the minimum number of ratifications has not yet been reached.

We are meeting here in Africa at a time when the community of nations is trying hard at the United Nations General Assembly in New York to chart a course to help Africa break out of its current economic and social crisis. As an organization with its headquarters in Africa, UNEP is especially sensitive to the damage inflicted by drought. In many countries it appears as if the social, economic and environmental fabric is being torn apart. At such a time, countries may be excused for not thinking of the long term, but never was such a course more needed. Severe as it may be, this drought is not without precedent. Our collective task in the United Nations is to help governments prepare for the next. We pray that the worst of it is over, and there are signs that it may be. But we know, as sure as night follows day, that drought will strike again.

As the international community responds with food and other forms of emergency relief, we must not deceive ourselves that it provides a lasting solution: it is 'Band-Aid' development. It can only help heal the wound, it cannot stop another from being inflicted. Our challenge is to give a future to the herder who has seen his cattle perish, a future to the farmer who has been forced to eat his seed-grain, and a future to the women who keep vigil around the village well.

Few economic prescriptions have taken into account the biological dimension of Africa's social and economic crisis and the world economic depression in general. There can be no recovery worthy of the name unless nations take steps to conserve their croplands, grasslands, forests, water systems and fisheries.

UNEP has long recognized that one of the most formidable constraints to effective environmental management is the view that environment is at best a supplement to development, at worst a barrier. There are signs that the sectoral approach is beginning to break down, but we still have far to go.

No one is better placed than parliamentarians to end this litany. Parliamentarians can provide the impetus at the legislative level to channel growth in developing countries towards sustainable growth. You should persist in reminding the executive that the environment is not a sectoral interest to be swept aside: it is an integral part of any sound development. I hope the parliamentarians from developed countries participating in this conference will take back to their governments the message from their Third World colleagues that the poorer nations are in a position to learn from earlier mistakes, and that they seek development without destruction. They should insist that environmental concerns are fully integrated into development assistance packages. To the parliamentarians from the developing countries, I wish to say that the traditional respect shown for the environment by almost every culture in your part of the world must be revived where it has been lost, and nurtured in those countries where traditions remain strong.

We must work together to put aside the axioms of the past, axioms that assumed infinite resources, there only to be plundered. We must recognize that our life-support systems are already strained to the limit. Apposite and tragic examples such as the African crisis are the harbingers of similar ecosystem collapses in fisheries, flora and fauna. I will venture at this stage to identify a few goals which you, as parliamentarians, may wish to consider when debating various agenda items. These include:

1. your role in formulating environmental legislation that is workable and has relevance to legislation on broader social and economic matters;
2. your role *vis-à-vis* the executive: to ensure that development activities are sustainable by being environmentally sound, and to assist them in implementing strategies to achieve this crucial goal;
3. your role *vis-à-vis* your constituencies: they have shown concern, but for that concern to be moulded by the executive into constructive and coherent strategies, you need to listen to your constituents, talk to them and lead them

into meaningful participation in decision making and, more importantly, into the implementation of environmentally sound programmes;

4.   your international role, which includes working to build better relations with neighbouring states, to promote international cooperation, to show commitment to the interdependence of nations, however remote, on environmental issues, and to ensure your countries' political and financial support to international environmentally oriented organizations such as UNEP; you have an important role to play in diverting some of the colossal resources – human, natural and financial – poured into preparations for war and destruction and channelling them into the prevention and abatement of environmental disasters.

Environmental management is a question of survival in more ways than one. As we destroy basic resources, we run the risk of aggravating already tense political situations. If war over oil is conceivable, then why not war over water? World-wide, 200 river systems are shared by two or more states, and already we have seen how pollution and unsustainable withdrawals have created friction. The pressure will grow. Using current technology, industry will require four times more water by the year 2000 to keep the same number employed as in 1975. Food shortages and famine certainly require both much more and better agricultural development. This will mean escalating demand for water. Aspirations for better standards of living, especially in developing countries, lead to another mounting claim on water for domestic use. Looking ahead to the coming years, this troublesome thread which looms in almost every aspect of natural resource use threatens to come loose and unravel our achievements of the past years. This possible connection between environmental issues and issues of national and global security underscores the importance of our working together, and UNEP will be honoured to walk by your side in this endeavour.

26

Protecting the ozone layer

Statement to the Opening Ceremony of the Conference on a
Convention for the Protection of the Ozone Layer

Vienna, Austria, March 1985

The Global Convention on the Protection of the Ozone Layer is the latest in a series
of efforts undertaken or supported by the United Nations Environment Programme,
aimed at reaching broad agreement between countries on environmental problems
and their solutions. Since the inception of UNEP, it has been clear to us that, because
so many environmental problems are intrinsically international, their solutions will
be arrived at when agreement likewise crosses borders. Over the years we have
learned that no aspect of our business is more valuable than international accord,
nor has any aspect been so satisfying to witness as when countries set aside political
and economic differences in order to agree upon environmental measures that have
been shown to be mutually beneficial.

UNEP's Regional Seas Programme has seen some of the most positive examples of
cooperation between countries experiencing serious difficulties on other fronts, some
representatives of which have met at the environment negotiating table while their
governments were in an actual state of war.

But this global convention differs from all other international conventions in
UNEP's history. There is nothing regional about this issue. The ozone layers protects
every square metre of our planet and therefore every person from every continent
and country. The atmosphere that we hope to save from potentially irreversible
imbalance makes up an entire component of our environment – it is as if we hoped,
with one agreement, to prevent permanent damage to all the oceans or all the soils in
the world.

Such is the magnitude of the challenge we faced when UNEP's Governing Council
decided, more that 10 years ago, to initiate work on establishing a plan of action to
protect the ozone layer and on a framework convention that has culminated in this
conference.

Praise is due to everyone involved in this delicate and crucial project: to the
Working Group which met seven times in three years to formulate the framework
convention and the protocol; to the Coordinating Committee on the Ozone Layer
for its essential contribution of scientific facts and essential data; to the World
Meteorological Organization for its scientific assessments and cooperation with
UNEP; and to all of the NGOs, scientists, and industries for their years of hard work
in putting forth ideas and theories that contributed greatly to the progress of the
Working Group.

I mentioned two of the differences that set a global ozone convention apart from
others in the environmental field. But the most important difference of all is
responsible for most of the difficulties we have had in deciding the most reasonable

and equitable response to the threat of a modified ozone layer. It is that depletion of the ozone layer is not yet upon us. In previous conventions, such as the Convention on International Trade in Endangered Species, or the Regional Seas accords, we acted only after the threat to certain species was already evident (and in fact after some species had already disappeard forever), or after pollution threatened to make coastal waters unsuitable for fishing or bathing.

But this time we face a distant, amorphous threat. This is the first global convention to address an issue that, for the time being, seems far in the future and is of unknown proportions. This convention seems to me the essence of the 'anticipatory' response so many environmental issues call for – to deal with the threat of the problem before we have to deal with the problem itself. As we all realize, in the case of the ozone layer, facing the problem itself might already be too late. That we are taking the anticipatory approach is a sign of a political maturity that has developed over the years, which recognizes how vital it is that we act to prevent environmental degradation or disaster with wisdom and foresight.

There is no doubt that a depleted ozone layer would harm people. But if we are lucky enough not to have laid the trap of a serious depletion already, which is by no means certain, then the people affected are not likely to be those of us in this room. We act now for the future. Those who could be threatened are the future generations that will have to live in a world that, through errors in judgement or mere short-sightedness, we risk making uninhabitable.

Our great frustration in ensuring that such errors are not made has been that our understanding of the problem has changed rapidly, even from one Working Group meeting to the next. Not everyone agrees on any one single theory; and yet we are acting now because we realize that to experiment with the make-up of our atmosphere is to experiment with the health of all humans and the welfare of all the organisms upon which we depend for survival.

Our understanding of the effects of increased UV-B radiation has also changed: the concern now is that a 1% ozone decrease results in a 2% increase in UV-B radiation, and as much as a four-fold increase in certain skin cancers and other biological effects. Further worry is the subject of continuing debate: even if ozone were to increase over the next several decades, what would be the climatic ramifications of the vertical redistribution of ozone that would accompany such an increase?

This and other questions are still being investigated – all the more reason to act, and to act now. The plethora of scenarios to come out of the last decade of research does not provide any excuse for complacency. We have seen come and go the near-doomsday scenarios of the 1970s, followed by revisionist theories that left room for optimism, and very recently a return to what could easily be called a doomsday scenario, the aptly-named theory of 'chlorine catastrophe'.

I can only say for certain that unfounded optimism has no place in a matter as serious as this one. From a scientific and environmental standpoint, we cannot afford to hope that a particular scenario is the definitive word on ozone depletion – we must prove it.

But while we wait for that proof, I cannot see the rationality of an argument which insists, in the face of scientific uncertainty and disagreement, that because no statistically significant depletion of total ozone has yet been recorded, we can go on producing chemicals that affect the atmosphere in unknown ways. If the long years of research have shown anything for certain, it is that we do not know enough about the interacting components of our atmosphere to take even the slightest risk of ruining its fragile balance. If our efforts here prove to have been unnecessary at some later date, in light of new and definitive scientific evidence, still our time will not have been wasted. At least we will retain the pride of knowing that we did not gamble with the only protection from UV-B radiation we have.

Fortunately, the first steps in reducing production of chlorofluoromethanes (CFCs) were easy enough to bear because a good proportion of their uses were frivolous in comparison to the perceived risk to the ozone layer at the time. In my opinion, in order to go far enough that we err on the side of caution in this issue of unprecedented importance, it is essential that governments not only ratify the instruments and texts agreed at this meeting, but also that they adopt protocols that are commensurate with a prudent interpretation of the available facts.

Preventing future calamity requires not only agreement but action. Governments and other responsible groups are usually accused of reacting to crises rather than foreseeing and preventing them. We have an opportunity here to show that experts, scientists, lawyers and governments can foresee potentially catastrophic dangers, and prevent them from happening.

Once accord is reached here in Vienna, as I am confident it will, parties to the convention must take specific measures to see that the spirit, as well as the letter, of the convention and protocol is acted upon. Only then will we have succeeded in allaying the world's understandable fears about the depletion of the ozone layer.

In closing, I can only repeat that, if there is an environmental problem for which tardy response is absolutely unacceptable, it is the possible threat to the ozone layer. It is hard enough to cope with the permanent disappearance of a species, or the death of a lake, or the turning of fertile lands into desert; but in the case of ozone depletion, who could forgive us if we reacted too late?

27

Global concern about environment

Opening Statement to the Thirteenth Session of the Governing Council of UNEP

Nairobi, Kenya, May 1985

The message of UNEP's 1984 State of the Environment Report was that environment crucially affects and, in turn, is affected by social and economic development. We produced that document before the dimensions of the African crisis had become widely known. As Africa's tragedy has developed, that thesis has been underscored, and now a world-wide public has begun to appreciate the environmental dimension of this continental crisis. The public has been shocked, not only by visions of wasted people, wasted bodies of men too weak to move, of mothers too weak to nurse their young, and of children too weak to cry, but also by scenes of wasted landscapes.

The sense of shock has been shared by the world's leaders. VIPs visiting Ethiopia, for example, have come away with a much deeper understanding of Africa's environmental crisis. Typical was a remark by the Vice-President of the USA, Mr George Bush, who said he had seen the terrible results of 'mismanagement and ecological disaster'.

The outpouring of international concern has been remarkable. But now the public, which has responded with such magnificent generosity, is beginning to wonder if next year, and the year after, famine will continue. Their concern is justified. When the rains finally come, they will wash away still more topsoil and thereby create a new generation of environmental refugees in search of foreign aid. Our common humanity will prompt us once again to act, to rush food and grain from the USA, Australia and Europe and financial aid from those who can spare it. It will be called 'drought' and 'famine' and 'natural disaster' and we will be revisited by the ghastly pictures that have moved the world to pity and sacrifice.

The public do not need experts to tell them that land is being degraded to a point where it cannot support the people, and they are asking: did we learn nothing from the last drought? Why must we keep lurching from crisis to crisis? I am asking the same questions. Why do we always have to wait for a disaster like a famine or the Bhopal and Mexico City incidents before taking action?

As the beleaguered inhabitants of Sudano-Sahelian Africa know only too well, the crisis precipitated by the drought of the late 1960s and early 1970s never went away. Numerous reports by the UN and other organizations warned that another famine was likely. In 1983 I stated before the Second Committee of the General Assembly that the drought was 'threatening to bring starvation and suffering on an unprecedented scale'. But now is not the time for recrimination. We should be grateful, at least, that the outside world is now more aware of the scale and intractible nature of the crisis.

It is against this background that we must ask if the response is all that it might be. Food for the starving as a top priority, yes, but, as I stated during the last session of the General Assembly, 'we should not deceive ourselves that this will provide a lasting solution – it might help heal the wound but not stop another being inflicted'. Only more assistance to help governments establish better planning in the medium and long term can achieve this goal. The same principle applies to chemical disaster. More stringent safety measures at chemical plants are required, but also more controls over the manufacture and trade of pesticides, and more education about their use. According to the World Health Organization, each year, in numerous small-scale tragedies, five times more people are killed by pesticide misuse than perished at Bhopal, and over a million people are poisoned.

Now, while questions are being asked about the fundamental direction of development, I believe the international community has an opportunity to learn from the mistakes, and to begin the task of redirecting economic development to securing the resource base. The African crisis will not go away, even if, as we pray it will, a period of sustained rainfall returns. With only 19% of Africa suitable for agriculture; with per capita food production falling; with population in a large number of its countries set to double over the next 25 years; with projections that real income will fall over the next 10 years; with a deficit in the balance of payments of US$21 billion in the three years from 1981 to 1983; with anticipated further deterioration of the terms of trade by 4–5% in 1985; with only a very limited possibility of increasing exports; and with very low expectation of increasing national savings – the outlook is bleak. Yet there are signs of hope, even in the worst affected countries.

Helping nations build up indigenous environmental, analytical and management capacities must be the first priority. UNEP is strategically placed to assist countries in tackling their most serious environmental problems. We do not pretend that environmentally sound projects will, by themselves, solve Africa's crisis, but they can be a spur to positive action elsewhere.

If the international community had taken note of our consistent warnings that the effects of natural disasters like a prolonged drought could be mitigated, the cost, according to a Red Cross estimate, of the current emergency relief effort could have been cut by 60%. Money going to saving lives might now be helping people to help themselves. This is what is happening in a small area in Southern Ethiopia which is escaping the worst ravages of the drought because a series of water and soil conservation projects mounted by non-governmental development groups and the Canadian Government have succeeded in enlisting the involvement and enthusiasm of villagers. In a Sahelian country, a treeplanting and windbreak project has boosted millet production by 40%. These are the 'signs of hope', the success stories which can breed more success.

With so much attention focused on Africa, we should not forget that in other developing regions environmental destruction also raises the spectre of man-made disaster. Africa is only at the sharp end; a cautionary warning of what could happen elsewhere – in Bangladesh, in Haiti or other desperately poor countries. Even in the better off developing nations, the situation is becoming serious. In nothern Thailand, for example, watershed destruction is now so extensive that one governor has warned that his province could become 'another Ethiopia within ten years'.

The most forbidding challenge facing the global community is feeding a projected turn-of-the-century world population of over six billion people. It can be done. In this year's State of the Environment Report, we are suggesting how we can properly apply environmental management for sound development to ensure that no one should be malnourished by the year 2000. There are sufficient natural resources, technologies, expertise and other resources to create a decent quality of life for these six billion people, provided nations use these resources properly and adopt sensible agricultural, pricing and marketing policies. New agricultural technologies and the rapid advances in genetic engineering of crops present us with opportunities to boost production on a sustainable basis. Rational population policies being pursued now will not halt this century's population surge, but they will leave the community of nations better prepared to satisfy the needs of people in the next century. This is the kind of time scale which we should be planning for.

We are now correcting mistakes made a decade and more ago, and the tragedy is that those mistakes are still being made. Inertia in decision-making circles ensures the survival of bad habits. In governments, major decisions about the fate of our environment are taken in the big-spending ministries. On the whole, environmental machineries still play a junior role in development planning.

More resources, more manpower and more money need to be channelled to strengthen these machineries. Even were these to be made available, they are not the whole, or even the major solution. What needs to be done is to increase dramatically the environmental awareness and competence of the energy, industry, agriculture and other powerful government machineries. These are vast but neglected fields for bilateral and multilateral cooperation.

Equally important is the need to improve coordination among these decision-making bodies. The setting up of National Environment Committees, to include also representatives of NGOs, the scientific community, media, business and industry, could help ensure that nations pursue a genuinely multisectoral response.

Bhopal, Mexico and the African crisis have also demonstrated both the power and the limitation of the media. The images of the dead, the dying and the afflicted will fade, but the causes of the suffering remain. We badly need the help of prominent public figures to keep these issues before the public. As part of our response, UNEP is inviting people in the public eye to be Goodwill Ambassadors, not for UNEP, but for the environment. The response to our initial approaches has been favourable.

Not all the signs are negative. We can be encouraged that some of the main actors in the development process are beginning to appreciate that action to conserve our environment and improve natural resource management can provide a spur to growth, and that closer internatonal and regional cooperation is the only course to follow.

In late 1984 and early 1985, at three major UNEP-sponsored conferences, we struck a remarkable consensus with parliamentarians, with non-governmental development groups, and with business and industry. At the NGO forum and at the Inter-Parliamentary Conference on Environment – both held in Nairobi – we were in total agreement on the complementarity of environment and development. Programmes were agreed which will provide a common framework for moving our ideas into action.

In the UN and other fora, we have been directing too much attention to governments, when 75% of the world's development is financed by the private sector. At the Versailles World Industry Conference on Environmental Management, UNEP made a start at redressing that imbalance. A programme was agreed which aims to involve industry at all levels, from the shop floor to the boardroom. Among the many beneficial results was a decision to establish follow-up government and industry groupings at, respectively, ministerial and chief executive level. The first joint meeting will take place in November.

This means that our view that 'environment' is not and should not be the exclusive preserve of organizations like UNEP is now accepted. It also means that UNEP and its Governing Council are moving to centre stage. We have a new constituency which is looking to this organization to indicate how we can help manage for renewal. In this respect you will recall my suggestion to the last Council that, via the clearing-house, we should seek to set up international centres for applied studies and training in areas of environmental and natural resources management. I am pleased to report that the Swedish Government has expressed a positive interest in establishing an international centre for environment and development at the Swedish Agricultural University of Uppsala. I hope this development will lead to the setting up of more such centres.

We could be charged too that we have not given sufficient attention to youth and women. We are attempting to correct that imbalance as well. The International Year of Youth has provided UNEP with a starting point to open up a dialogue with young people. They are the next generation of decision makers, with the greatest stake in how well and how fairly we manage the earth's resources. We are embarking on three initiatives: asking young people to contribute to a 'youth environmental agenda'; catalysing a network of children's magazines to be distributed to schools; and establishing 'environmental volunteers'.

This year, 1985, marks the conclusion of the UN Decade for Women. A conference will be held here in Nairobi in July to review and appraise its achievements. UNEP intends to make a major contribution. Our main point is that the burden of the environmental crisis, especially in the developing countries, falls on women. Experience has shown that grassroots development projects will fail unless they pay full attention to the special needs and rights of women. Let us not forget that women produce more than half of Africa's food. Some grassroots movements – most notably the peace campaigns in Europe, the Himalayan villagers' tree-saving drive, the green-belt movement in Kenya and improved cooking stove programmes in the Sahel – have all been started by women.

Avoiding future disasters, planning development which will secure the natural resource base, is the uniting theme of all the proposals before this Council. I should like to highlight just three initiatives. Though directed at different sectors of our environment, they share features characteristic of all our proposals.

1. First is anticipation. The Ozone Convention agreed in Vienna in March is a testament to nations' political maturity because, for the first time, nations are agreeing to take steps to ward off a threat that may be still far in the future.
2. Second is assistance to decision makers. GRID (the Global Resources Information Database) is UNEP's new data management service which will enable environmental data to be transformed into usable information, essential for sound development planning. I am counting on the full cooperation of the federal government of Switzerland, the government of the Geneva Canton and the authorities of Geneva University to ensure that the GRID pilot effort starts operating in July and will become fully operational two months later.
3. Third is regional cooperation to protect shared resources. UNEP, in concert with the Zambezi River states and other UN agencies, is starting a Zambezi Action Plan. This is a development of profound importance.

The clear message from these and, indeed, all of UNEP's activities is that we must continue to concentrate our resources on bringing nations together to combat their indivisible environment problems.

Earlier this year UNEP convened a meeting of senior staff and advisers to consider our longer-term purposes and strategies. We reached a broad measure of agreement that UNEP must make some difficult decisions on which areas to concentrate upon. It was felt that UNEP should build in greater flexibility to respond to emerging issues and environmental emergencies, and should enhance its role as an 'honest broker' to bring together governments and UN agencies.

The General Assembly Resolution 2997 which established UNEP decided that one of the main functions of this Council should be to provide policy guidance to promote international cooperation and coordination of environmental programmes. We are, as usual, seeking guidance of a very high order, and to you falls the task of deciding how our minimal resources can be used to greatest effect. UNEP's role will not shrink but grow in importance, of that much we can be certain.

Any reaffirmation of our catalytic and coordinating function will require:

- a clear identification of how UNEP interventions can be made more effective;
- a determination of how the UN's environmental mechanisms can be improved;
- the setting of defined targets for UNEP, specific objectives for the UN system, and agreement on goals at the governmental level.

At the Session of Special Character, I told the Governing Council:

In 1982, nations have two choices: to carry on as they are and face by the turn of the century an environmental catastrophe . . . or to begin now in earnest a cooperative effort to use the world's resources rationally and fairly. [I also said then:] I would not be addressing this meeting today, nor would I continue to act as UNEP Executive Director, if I did not believe that governments – singly and collectively – will respond during the next years in a more serious way.

Three years later I am beginning to doubt if I was justified in my belief.

The most telling signal of how seriously governments are prepared to take our mission is how much they are willing to contribute towards it. The shortfall of US$25 million in contributions to the Environment Fund during the present biennium as against the resources envisaged by the Governing Council has had serious implications for the implementation of UNEP's mandate and of its approved programme. At a time of severe financial stringency it would, I admit, not be reasonable to expect a substantial increase, but I would ask delegates to consider how UNEP's activities could be stepped up through increased contributions to the Environment Fund, through support to particular activities using Trust Funds or Counterpart contributions, and through other innovative means. I am encouraged by the growing interest in our clearing-house operations. I hope this interest will soon materialize into solid and long-term cooperation with developing countries in dealing with their serious environmental problems. On my side too, I shall continue reviewing the structure of the Secretariat and our mechanisms of operation in order to make sure that we run our affairs as efficiently and as economically as possible.

The theme chosen for the 40th anniversary of the United Nations is the 'United Nations for a Better World'. A precondition for a better world is a better environment. While we can point to some improvements in environmental quality, to a gathering interest and concern for the environment among the public and decision makers alike, the overwhelming evidence is that the quality of the human environment is deteriorating. As pressure on the ecological foundations of the global economy mounts, so just as surely will the economic and political security of nations be undermined.

The forces that imperil this planet's imperfect and fragile peace are great enough, without adding competition for diminishing resources. On peace – its preservation and enhancement – rests the ultimate case for environmental action. I make no apology for stating what the cynic might dismiss as a platitude. Water, forests, fisheries, grazing and croplands are resources no longer taken for granted. Their exhaustion can lead to war, their enhancement can create unity and a sense of common purpose among nations where none existed before.

The need for strengthened international cooperation has never been greater. But there are many trends that are deeply disturbing. Foremost is a retreat from multilateralism. Failure to anticipate the famine, and failure to spread equitably the benefits of economic growth has sapped confidence in the effectiveness of international action.

Unfairly, I believe, the United Nations has come in for some overly harsh criticism. Its public image leaves much to be desired. This is worrying. Clearly UNEP, together with its sister agencies, must do more to inform the public of its many achievements, past and present. The UN's 40th anniversary commemoration therefore provides our organization with an important opportunity. I hope this Governing Council will decide to send to the commemorative session of the General Assembly a message presenting the achievements of the United Nations in the field of the environment.

Despite the setbacks, we cannot afford to lose our belief in the ability of the UN and other international organizations to promote multilateral action. The UN was established in the aftermath of one world war to prevent another. We therefore owe it to ourselves and to generations to come to keep faith with its founding fathers. We can build a better environment and thereby a better world.

28

Management of hazardous waste

Statement to the International Symposium on
Industrial and Hazardous Waste

Alexandria, Egypt, June 1985

Wasteful production processes are a folly that our resource-hungry world cannot afford. More than an extravagance, they are a menace to development. Wastes need not be accepted as a necessary curse, when human ingenuity has proved that waste-producing processes and practices can be minimized or avoided altogether.

There is no argument now: we all understand that the first priority in waste management is to reduce waste generation at its source, using other raw materials or processing them in a different way, often at a profit.

Two years ago, UNEP and the World Health Organization sponsored the publication of *Policy Guidelines and Code of Practice* on the management of hazardous wastes. It is a valuable, authoritative text and remains useful two years after its publication. But on one detail I no longer subscribe to its guidelines. According to the report, 'unfortunately, the wide-spread adoption of waste reduction techniques by the manufacturing industry is likely to occur only where economic advantages are to be gained'. We know now that we should have deleted the word 'unfortunately' from the text. Since those guidelines were published, we have developed an airtight case to show that ecology and economics are mutually reinforcing. As I said, there is no argument now.

At the World Industry Conference on Environmental Management (WICEM) held last November at Versailles, hundreds of leaders from industry, government and non-government organizations, along with parliamentarians and representatives of labour and the scientific community, considered the case and reached consensus that economic growth is compatible with environmental protection, and that an anticipatory and preventive approach is the first priority. Practical follow-ups to WICEM are planned in various fields, including waste disposal.

Our position in UNEP is clear with respect to waste in general. When the processes we adopt do produce waste, we must choose those which minimize the ill-effects – the minimum risk to the safety of people and the environment, the minimum difficulties in handling, transport and disposal.

Waste that we cannot avoid producing may be reused by others. It can be exchanged in ways profitable to both sides. Whenever we legislate and regulate waste management, let us carefully take steps to promote such an exchange.

Since the economic value of residue is nearly always quite small, there is normally a limit to the distance that can separate the waste producer and the potential user. For example, if the cost of treatment and safe disposal of one ton of waste in Western Europe is 10, 20, 40 or US$60, it is most unlikely that a profit of over US$100/ton will be made if it is received in, say, Djibouti, considering the cost of transport and handling. That is, unless it is an extremely hazardous and difficult waste to handle and store. In fact, UNEP gave very unfavourable comments on such a project to the UNEP representative in Djibouti last February. Our policy is to oppose the transfrontier transportation of environmentally hazardous wastes for the purpose of ultimate disposal. We believe that the problem should be solved in the country generating the waste produce.

Waste that cannot be avoided, re-used or recycled must be disposed of in an environmentally acceptable way. Of the thousands of millions of tons of industrial waste being created every year, it has been estimated that a significant fraction – up to 2.5% in the European Community – can present substantial environmental hazards: toxicity to people (especially workers) and to animals; risk of fire or explosion; long-term environmental hazards or chronic toxicity upon repeated exposure; and the potential to pollute drinking water. We have four difficult issues on our own agenda.

First hazardous waste must be clearly defined. Its composition, physical form, quantity and long-term threat must be determined. As specialists, you are aware of the diversity of approaches that exist in defining a hazardous waste. Should we list criteria for environmentally harmless wastes, and define all others as hazardous, or should we produce inclusive listings of hazardous wastes? How should we characterize wastes that are hazardous in only part of their management cycle? These are the kinds of decisions upon which effective management depends. We are relying upon you, the experts, to come up with imaginative answers that will make our job easier.

The second issue is the choice and development of waste treatment and disposal technologies. This is also dependent on a large body of knowledge and experience and, in this regard, UNEP is encouraging further investigations of incineration technologies. But it is a fact that policy on treatment and disposal technologies varies widely from place to place. Take land disposal, the major management method for waste in many countries. Problems have arisen at former waste disposal sites in a number of industrialized countries. Toxic scares have intensified public opposition to waste disposal sites, making it ever more difficult and costly to open new ones. Again, approaches vary, and we rely on the experts to guide us to the final decision. Should we 'concentrate and contain' absolutely, dealing with materials and waste leachate, or should we aim for progressive and ideally controlled 'dilution and dispersion'.

Third, thanks to efforts such as yours, and to publications of UNEP's Industry and Environment Office and International Register of Potentially Toxic Chemicals, we are getting to know more about hazardous waste problems in developing countries, although very little compared to the lessons already learned in the more industrialized countries. Universally, uncontrolled or open dumping is an unsatisfactory method of disposal for hazardous waste, not least because scavenging is a widespread practice. An urgent priority in many countries will be the elimination of open dumps. But this action will often only be possible when appropriate technologies are available which offer significant environmental improvements at a reasonable cost. The 'best available technology' might potentially reduce the risks by a factor of 100, but at a cost which cannot be met in developing countries. The overall result would be that nothing is done. The 'best practicable means', on the other hand, might be a less sophisticated methodology which will reduce risks by a factor of perhaps 10 at a fairly modest cost. The real world demands trade-offs and we must come to terms with that fact.

The fourth issue is that not enough is known about the basic characteristics of the hazardous waste problem in developing countries, whether with regard to the nature and geographical distribution of the waste, or with regard to the effect of extreme heat and high humidity or floods on disposal options. Research and development work are urgently required to define the nature of the problem and to develop appropriate technologies. Another area of concern is the handling of existing hazardous waste-containing open dumps. Can these dumps be closed without posing problems in the future? For like so many environmental and political problems facing the world today, land disposal of wastes involves solidarity between generations.

Before I conclude, let me propose 10 suggestions which you may wish to consider to see if they would be of any help to governments and industry in improving the management of their industrial waste:

1. More information exchange of case studies on successful waste disposal practices is particularly valuable. Without such an exchange of information, successes and failures can go unnoticed, and valuable lessons go unlearned.
2. Early collaboration between industry, local authorities and national administrations during the planning stage is vital, not least in the selection of sites with best access to treatment and disposal facilities.
3. The possibility of treating and disposing of municipal and industrial waste jointly must always be considered.
4. Providing encouragements and incentives for low-waste industrial processes and technologies, and for adhering to regulations, is also crucial.
5. A definition of the best technology, and best practicable means, and an analysis of cost-effectiveness of alternatives, will greatly aid in the decision every government and every affected industry will have to make.
6. Industry staff, at all levels, must be trained and informed, so that environmental aspects of industrial waste management are considered as a matter of course.
7. Similarly, lessons gained by industry on waste management must be brought to the attention of the relevant governmental authorities charged with enforcement.
8. The labour force and general public must not be kept in the dark about an issue which can arouse fear and suspicion.
9. The dumping of wastes at sea, and the export of hazardous wastes to other countries, must be prohibited.
10. When assessing the costs of the clean-up of wastes, the benefit to the welfare of the local population and to the environment must be taken into account.

The feeling at the World Industry Conference on Environmental Management was that we, the theorists of the environment, were extremely pleased to be communicating with the practitioners. It was found at that conference that there was no need to search for common ground: it already existed. Industry cannot afford lax methods of waste disposal any more than the environment can. And so we turn to you, the experts, to ensure, through your research and experience, a future free of the threats posed by hazardous wastes.

29

Women and the Earth's traditions

Statement to the World Conference on Women

Nairobi, Kenya, July 1985

We come together in Nairobi at a time of tragedy for Africa. Many of you have witnessed first hand, or in pictures and words, the terrible suffering of your fellow human beings.

Today, the world struggles to save the lives of millions of refugees from denuded lands. Many voices warn that immediate aid is not enough, that we have to attack the roots of the tragedy lest we risk perpetuating and widening the cycle of famine and resource destruction.

The plight of Africa has been labelled 'drought' and 'famine' and 'natural disaster', but at the bottom it is an environmental disaster and unless we alter how people think about Africa and its plight we will be revisited by the grim visions that have stirred the world's conscience and compassion. The lesson we are learning from this African crisis is that we should be looking for sounder alternative patterns of development, new patterns which use the natural resources of the earth – our land, our trees, our soil, our air, our water – more rationally for the benefit of humankind.

Solutions are at hand. Two themes of this Conference – development and peace – are the first concerns of UNEP. In the environmental movement, we have long recognized that the hardships of people will only be alleviated by development. Development and human welfare are inseparable. Development means agricultural technology, industry, housing, and production of all kinds that will supply jobs and amenities for growing numbers of people. But poorly executed development can also lead to overcropping of lands, overdependence on cash crops, pollution of water and air, deforestation, desertification and the squandering of energy. Therefore, our constant emphasis, no matter what the environmental context, has been on 'sustainable development', or development that will preserve, not exhaust, the resources on which it feeds.

We have come far enough in the short, 12-year history of the United Nations Environment Programme to believe that we have at least some of the answers, that we have workable methods of communicating those answers, and that we have successes and failures that can guide us in the future. What we lack, sadly, is a manifestation of the political will and foresight that is the key to seeing the concept of sustainable development affixed in the minds and entrenched in the policies of those who lead us.

I look to you, women from all walks of life, to join us in defining and, most crucial, redirecting the course of development to prevent further environmental catastrophes. The burden of environmental degradation and crises has always fallen, and continues to fall, on women, especially in the developing countries. I am speaking of the women in Africa who labour long and hard in the fields to produce more than half the food for the continent; I am speaking of the women of the Sahel who keep a night-long vigil by their wells to glean a few drops of water; and I am speaking of the women of the Himalayas who, it has been reported, have been driven to suicide because they could no longer tolerate the burden of the firewood crisis.

Women who are in positions of influence have a special duty to represent those at the sharp end of the environmental crisis. More and more women now occupy positions of influence. Women are teachers, lawyers, engineers, parliamentarians, heads of state, scientists, law-makers, industrialists – and environmentalists. Women are, by tradition, managers of finite resources, whether it is the limits prescribed by a pay cheque or the limits of firewood, food and water to meet family needs.

We can simply look into the workings of our private lives and the structure of families and communities, down through the centuries, to shed some light on how you have nurtured finite resources. Woman's traditional role as manager of the individual household, is a script for how we should better manage our global household.

I am not speaking of the role of women as a fixed one in modern society: their social and political equality is long overdue and sadly, still has far to go. I am speaking, rather, of an awareness born of centuries of experience. Perhaps it is this tradition that has led to what many perceive as a greater awareness on the part of women of the need to preserve, protect, and equitably share the dwindling resources of our global environment. In diverse circumstances, women have often been the first to lead the protests against chemical, water and air pollution, and they have been leaders in environmental education and citizen action leading to political action.

This leadership is highlighted by two important environmental initiatives, both initiated by women in the developing world. One of these, the Green Belt Movement, is a project of the Council of Women of Kenya. Kenya is a country where 90% of the population live in rural areas, and where more than 85% of energy comes from fuelwood. It is the women of Kenya – and every other African country – who are the firewood collectors: they are therefore in the front line against soil erosion. The Green Belt Movement is a grass-roots, tree-planting movement with its base in local communities, where women and young people set up nurseries and then supervise the planting and care of the seedlings. There are more than 60 nurseries in Kenya already, producing millions of seedlings.

The 'Chipko' movement in the central Himalayas is another example of women in action to prevent irrevocable loss of their resources. Angered by the selling-off of their trees by the menfolk to timber companies, they took non-violent action to protect the hillside tree cover. Drawing their inspiration from the Bishnois women of Rajasthan who, centuries ago, were killed while protecting their trees, the Chipko movement has spread throughout the Himalayas. It is now a powerful force promoting the right of local people to retain control over their resources.

Restoring the balance between people and resources, and between environment and development, will take time, levels of concern and participation that do not yet exist. Restoring the forests and grasslands to their former pristine condition is no longer possible – the human population and its demands are too great. But we can achieve a balance between development and resource renewal. This is what we mean when we talk of solutions. I would draw your attention to UNEP's 1985 State of the Environment Report which indicates how this might be achieved. It is not enough to talk about preventing further damage; we must be prepared to shoulder the burden of healing the ailing earth where damage has already been done.

Those among you who have kept apace with developments in the environmental field will know that our planet is strained to the limit by abuse and degradation of resources. I could point to the prospect of a global population of 10.5 billion in little more than 100 years, and go on to describe the obvious strains of so many mouths to feed, so many crops to grow, so many homes to build, so much fuel and firewood to be consumed. Or I could state that acid rain has killed life in thousands of lakes in Scandinavia and Canada, while forests in Central Europe are being ravaged by acidic fall-out. Or I could note the tragedy of the world's freshwater systems: the term 'fresh' hardly applies to much of the world's drinking water supply, and two thirds of the world's rural population have no access to clean water. Or I could refer to the migration to the cities in the developing world where the new arrivals encounter cramped squalor, discrimination, new health hazards, not least from industrial catastrophe. We do not forget the Bhopal and Mexico City tragedies.

The human species has reached a stage which I would hesitate to call the point of no return, for we take heart in past successes even in the face of continuing degradation of the environment. But the facts are before us and are sufficiently alarming as to provoke dismay in the face of so much apathy, so much avarice and so much short-sightedness on the part of those who could lead us out of the environmental traps we have laid for ourselves.

The second theme of your Conference is Peace. As has often been said, peace means more than the absence of war. Is the world at peace when 40% of the world's population have no effective medical services? Is the world at peace when 3000 million people lack access to safe drinking water? Is the world at peace when 750 000 die each month from water-borne diseases? Is the world more secure for the fact that military expenditure will surpass 1000 billion dollars annually by the year 2000? Even now, military research and development absorbs scientific and technological resources ten times as great as those available to all the developing countries together.

We must take inspiration from the peace movements which refuse to accept the inevitability of the arms race. Significantly, it is women who are spearheading many of the campaigns; it is women who are saying this insanity must stop. To arm to the teeth and squander resources in preparation for wars without winners cannot represent meaningful security. In the mean time, environmental resources are squandered, resources which support human welfare and human health.

Women must use their hard-won responsibilities and social positions in order not to succumb to the *status quo* that has left the planet in such perilous circumstances. They must press for the kinds of national and international actions that will bring about a change for the better. No region of the planet is exempt from environmental bankruptcy. In developed countries, air and water pollution persist and new problems of chemical contamination loom as industry produces increasing quantities and varieties of new compounds.

What should be the role of women in Asia, and in Latin American regions experiencing unprecedented industrial growth? Women have a crucial role to play in fighting regressive terms of trade, punitive land tenure systems, wasteful consumption patterns and the other 'invisible' forces that cause environmental despoilation. Lamenting the loss of forests, pollution and so on is not good enough. You must turn your attention to the decision-making processes which cause that destruction. With your help we can turn environment into a major decision-making priority.

The objective is sustainable development. Shortages of water, food, wood and fuel are as much threats to stability and security as are political or ideological differences.

All these natural resource shortages can be overcome through environmentally sound development planning and implementation. You have to educate others and educate the generations that will inherit this earth. We are appealing to you to teach your children and your community that there are no permanent technological fixes in nature. We men have pretended for too long that we can conquer nature, when the most we can hope for is to live in harmony with the Earth, within nature's natural limits.

Every vocation carries with it environmental responsibilities – biology, engineering, agriculture, political science, or economics. Students of these disciplines must see the imperative of environmental considerations in the real world. The environment, our Earth, is not something to be conquered, it is something to be protected as we protect our own children. Women from all over the world can find, as a bridge to each other, the salvation of the Earth, whether in movements such as the peace campaigns, Chipko and Green Belt, or in the service of parliaments, corporations, universities and citizen groups. People, not governments, have brought us thus far in the environmental movement. It is people who, through protest or concerted action, compel industrial or agricultural development in their countries to consider the environmental consequences. Women, as more than one half of the human race, must mobilize. You who often suffer first and are consulted last; you who must live with the consequences of decisions without having the forum to voice your objectives. You must join with us.

In our search for workable solutions to these tangled issues, UNEP has seen the potential, as well as the limitations, of international actions on environmental issues. Whether it is the threat to the ozone layer, the protection of coastal waters, the conservation of endangered species or the responsibilities of industry in the preservation of resources, there is a growing awareness. But it is not enough, and it will not suffice if local residents do not have rights or a voice in their future: if they cannot say 'No' to development schemes that are short-sighted, overlooking environmental considerations; of if they cannot say 'No' to astronomical arms expenditure or to industries that callously disregard the effects of their pollution on distant lakes and forests.

The human commitment needed to heal the world – a world in which every citizen and every government plays a role in preserving the Earth for future generations – is surely not lacking among the women assembled here. The quest for women's rights, rights too often denied or suppressed, has brought you into the realm of politics, both global and regional, and has taught you that the *status quo* is not monolithic, and is susceptible to reason. Industry can be shown that prevention of environmental degradation pays, and pays now. Whole regions can be shown that cooperation on environmental issues can lead to cooperation on other fronts. The energy and sheer force of will that has brought the global women's movement so far so fast is desperately needed by the environment and development movement, and now, whether you work in local communities or in the corridors of conventional power. You must join each other to reclaim your future.

If there must be war, let it be war against environmental contamination, nuclear contamination, chemical contamination, against the bankruptcy of soil and water systems, against the driving of people away from their lands as environmental refugees. If there must be war, let it be against those who assault people and other forms of life by profiteering at the expense of nature's capacity to support life. If there must be war, let the weapons be your healing hands – the world's women, in defence of the environment. Let your call to battle be a song for the Earth.

30

An urgent role to save the Mediterranean

Statement to the Fourth Meeting of Contracting Parties to the Barcelona Convention

Genoa, Italy, September 1985

Our pioneering environmental agreement, the Mediterranean Action Plan, is now 10 years old. The anniversary provides us with a valuable opportunity to review not only successes and achievements, but also the disappointments and failures. It also gives us a welcome opportunity to plan for the next 10 years. This meeting is taking place at a critical point in the brief history of the Mediterranean Action Plan because, like the sea it is intended to protect, the Plan is not as healthy as it should be.

UNEP frequently cites the Mediterranean Action Plan as one of our foremost achievements. In very difficult conditions, a great deal has been achieved in a very short space of time. The Action Plan became a blueprint, not only for UNEP's other Regional Seas accords, but also for other international accords aimed at addressing a wide range of environmental problems. Progress with the implementation of the Action Plan became a yardstick by which we could judge other agreements.

On all levels – technical, scientific, legal and, above all, political – the Plan set a very high standard. Mediterranean states confirmed our belief that governments are prepared to put a shared concern for the environment beyond the reach of political divisions. Therefore, if I sound somewhat critical in my assessment of where we stand, it is that I am merely keeping faith with the high standards set by governments at Barcelona in agreeing the Convention for the Protection of the Mediterranean Against Pollution and later the protocols.

I wish to touch on some specific points. *First*, the Contracting Parties have created an apparatus which has not yet been put to full use. There is broad satisfaction with the Convention itself, the legal instruments have been given near universal approval, the quality of scientific information is high, a secretariat and an array of supporting units and institutions are in-place and working, and yet this elaborate organizational structure is in danger of being dismissed as a facade. It is as though all the goods are in the shop window and none on the shelves inside.

Certainly, progress has been achieved in the past 10 years. Most progress has been made with scientific monitoring, but we still do not know enough about the complex movement of the Mediterranean's water masses, in particular their role in transporting and distributing pollutants. This gap in knowledge needs to be filled, and urgently.

MED-POL, the Pollution Research and Monitoring Programme, is the foundation of the Action Plan. Through cooperation, an overall assessment of microbial pollution of beaches, of shellfish and shellfish-growing areas and of mercury pollution of seafood has been made possible. This is a major achievement by your scientists and your scientific institutions. But the goal of a basin-wide monitoring network, producing data on a regular basis, has not yet been achieved. Information on sources, levels, pathways and effects of pollutants provided by MED-POL's network of 80 capable national research institutions has not yet provided a solid basis for effective action on a number of other fronts. More resources should, therefore, go into improving the quantitative flow of information.

We can be encouraged that data yielded by MED-POL have been of the highest standard. The Programme certainly broke new ground in establishing the threat from land-based sources of pollution. But the 1978 survey, which formed the basis of the protocol on land-based sources, is in urgent need of updating. Since then, new generations of chemical pollutants have come into production. UNEP welcomed the speed with which the protocol was ratified, but we are disappointed with the apparent reluctance of many states to adopt the legally binding criteria proposed by UNEP, together with WHO and FAO. These criteria have not been challenged – no alternatives have been proposed – but no action has been taken. I hope that this meeting will decide to adopt them. If not these, then which ones? If not now, when?

Second, UNEP is especially concerned that the countries sharing a sea which can renew its waters only once over an 80-year cycle still permit an estimated 10 billion tonnes of domestic and industrial waste to be discharged each year. No sea, especially an enclosed sea, can accept with impunity such an onslaught. While we can state with reasonable certainty that the open sea is still relatively unaffected, many coastal regions are heavily polluted. As you know, coastal zones are the most productive part of the sea. At least one new protocol on offshore pollution needs to be urgently developed.

While scientific uncertainty may shroud the pathways and long-term effects of some pollutants, this is manifestly not the case with sewage. Governments at this meeting in Genoa should resolve to reduce this type of pollution, paving the way for its complete elimination – a task which is not beyond your technical know-how or your financial resources. While governments drag their feet on this issue, the public will remain sceptical about your commitment to protecting the sea.

For other pollutants, UNEP, although recognizing that further in-depth scientific research and monitoring is required, insists on immediate action to cut down gradually the serious industrial pollution. We welcome the new emphasis in MED-POL's second phase on systematic research, but this should certainly not be taken as an excuse for no action. It is quite evident that governments, in every corner of the world, when confronted with what is a seemingly costly control action or a politically sensitive regulation, fall back on the need for more monitoring and research. These are becoming smokescreens – excuses for inaction. If the Barcelona Convention is going to do any good at all it needs to move beyond monitoring and assessment. Until the Contracting Parties are willing to stand by their research and say 'yes, we now know enough to insist on regulation and action at the national level', the Mediterranean and the coastal areas around it will continue to be sick and will probably get sicker. Assessment is not an end in itself. It is a step towards policy formulation and planning and implementing required actions. There comes a time when we do know enough to call for regulation in certain areas. That time comes, and all too often nothing happens, and so we return to assessment.

I would like to remind governments that you entered into the Barcelona Convention to benefit from joint research and joint action. Each country is given the opportunity to use this sea more sustainably. The purpose of assessment has been to allow governments to build a stronger foundation for their economies. A dead sea – a polluted sea – and damaged coastal areas will benefit no one. Why are governments not acting on research findings? It is, I agree, always cheaper to do nothing now and hope that the problems will go away. The problems will not go away, and our assessments tell us that each year we wait, the more expensive our folly will become. Self-interest – an interest in the future – should be a spur for action and not an excuse for apathy.

Third, governments share our frustration over the delays and setbacks encountered with the Blue Plan and the Priority Actions Programme. Yet, the first phase of the Blue Plan has produced an interesting picture of this basin, shaped by a common history, sharing so many values, but as yet shy or slow in recognizing its common identity, and its common future.

In the final analysis the fate of the Mediterranean will be sealed by nations' development strategies. The need to follow strategies which pay full heed to the conservation of shared resources is critical. UNEP therefore urges states to take fully into account the alternative development scenarios which will develop from the Blue Plan and Priority Actions Programme exercise. In our view, special attention should be given to fields that are relatively undeveloped, such as environmental impact assessment, aquaculture and renewable energies.

Fourth, a great deal more effort too should go into establishing specially protected areas. We are particularly concerned to see that the Contracting Parties should take immediate steps to protect areas of special ecological significance and marine mammals threatened with extinction.

Fifth, despite overwhelming evidence that a genuinely multilateral response is the only effective means to protect the shared sea, the Action Plan has been seen too often as a self-contained programme. National development programmes on the whole have been formulated independently and have not reflected the priorities and commitments of the Action Plan.

Sixth, we have lost sight of the fact that the Action Plan was in large measure a response to the concern of the general public over the health of the Mediterranean. Unless a greater effort is made to keep the public convinced of the value of international action, support for the Plan will fade. The best way to maintain that support is to make clear demonstration of the practical benefits.

The catalytic role of UNEP in the Mediterranean is almost at an end. In all, we have contributed over US$8 million to the development of the Mediterranean Action Plan. An infrastructure has been created which matches the complexity of the subject it covers. But when we consider that the total support for the Action Plan through the Trust Fund amounts to US$4 million a year – granted there are more resources committed in kind by the national institutions involved – and when we consider that the purpose of the Plan is to protect a sea which nurtures over 200 million people, the financial support must be considered derisory. The present level of response in no way matches the magnitude of the problem. Control of land-based sources of pollution alone will require a new level of commitment involving governments, industry and local authorities in a basin-wide effort. A common peril requires a common response, not on paper but in practice.

Mediterranean governments do have the apparatus to bring their national and bilateral activities into line with the Action Plan. International obligations must be translated into national laws and practice; the provisions of the Convention should be implemented in their entirety; common standards should be adopted; national monitoring programmes must become operational in every coastal state; a regular flow of monitoring data must be assured; national contingency plans will have to be completed in every state; and the general public, through an improved public relations effort, must be kept better informed of the benefits.

In order to mark a common determination to nurture the fruits of the Plan, it would seem appropriate that you adopt a declaration that will allow this Plan to work for the people of the region. I have ventured to circulate to you a draft of such a declaration. I am also proposing that you adopt a programme of action with clearly defined goals for the next decade. In addition to accelerating the slowly moving elements of the Action Plan that I mentioned earlier, these goals should include the establishment of:

- reception facilities for oily residues in all major ports;
- treatment plants for sewer effluents in all cities of over 100 000 population;
- suitable outfalls for sewer effluents in all cities of over 10 000 population;
- measures to ensure full access by developing countries to technology and expertise in oceanography and pollution control.

If effectively applied, the declaration and the programme of action will put the Action Plan back on course.

On the occasion of the 10th anniversary of the Barcelona Convention, the Contracting Parties now have an opportunity to rededicate themselves to the objectives of the Mediterranean Action Plan, and to reconfirm their commitment to cooperative action.

You have the organizational structure, the know-how and, I believe, the political will to make the Plan work in the way the founding fathers at Barcelona intended. You have the historic opportunity to maintain the Mediterranean Action Plan's position as the foremost environmental agreement from which all others draw guidance and inspiration. I am sure you will use it.

31

Resource management – the human connection

Statement to the Sixth General Assembly of the Scientific Committee
on Problems of the Environment (SCOPE)

Washington, DC, USA, September 1985

SCOPE, as usual, is looking to the future: probing the options with characteristic
vigour. As an outsider I advance my opinions here with some hesitation, but as a
scientist standing somewhere between the worlds of science and of public affairs, I
think I can venture to share with you a few suggestions on how to bring the two
realms a bit closer.

SCOPE is a pioneer in establishing the scientific bases of environmental
management. My own belief is that any serious initiative to hold off environmental
degradation will have to start with a management perspective. If we could apply a
fraction of the scientific knowledge and technologies now available to us, the
practice of environmental management might become daily routine and the
environment might cease to be an issue for theoretical speculation and conjectures. I
claim that it is now possible to stop environmental degradation. It is possible, but it
is not happening. We have so many answers, but we do not apply them.

SCOPE produces a mass of valuable work, yet a lot of that work is not applied, or
used only on a small scale. If SCOPE could channel some of its effort into the
widespread application of the methodologies it developed for proper environmental
management, then we would all see the real fruits of its scientific excellence. Without
such redirection of scientific effort I doubt that we can make real progress in halting
the breakdown of various components of the environment in a large number of areas
around the world. By redirection, I mean the ability to weave scientific information
into the fabric of public decision making in an effort to ensure sustainable
development.

'Sustainable development' is a cliche in some circles. In others it means nothing. I
do not think we give enough thought to why its basic tenets are lost on many
decision makers, to why it is not working, and to how it can start working. I will
illustrate this with one example. Six million hectares of productive land are estimated
to be lost every year to the advancing desert. Another 21 million are also estimated
to be reduced annually to zero economic productivity. All this is equivalent to some
US$26 billion lost every year in agricultural production. To stop that loss by holding
back the desert encroachment would cost between two and three billion dollars
annually over what is being spent today for the next 20 years. To any manager this
would be a very profitable and attractive investment. But, unfortunately, the return
is not for tomorrow. It is for the medium term or perhaps even the longer term. It
will take some time before investors can reap the benefits of their savings.

Investors want results today and tomorrow, not promises for the medium and
long term. These are real constraints. The result is that almost nothing is done to

stop the trend. Our job is not to whistle in the wind and wish those constraints did not exist. Our goal should be to recognize the self-interest of governments, industry and special groups, and to show what can be done within those constraints. Our task should be to continue pointing out the links between lack of environmental management and recurring failure of development efforts: showing, for example, that famine is a painful symptom, but desertification and natural resource mismanagement are the causes of many natural shortages and Africa's recurrent pain. SCOPE can show how famines can be stopped through the introduction of better land-use practices. It can show even the USA that it could save the best part of a billion dollars a year in agricultural loss if steps were taken to avoid the compaction of soil. It can show that such saving could be repeated elsewhere. It can also show that almost all of our 'natural' disasters are exacerbated by poor management or non-management. SCOPE can also participate in demonstrating how sustainable development in the South is in the benign self-interest of the North, not only in terms of trade and resource flow, but also in terms of global climate and the balance of the bio-geographical cycles.

SCOPE and UNEP have been doing a substantial amount of work in these fields, but more is required. Without the ability to manage the land and the Earth's resources, a continent like Africa will be revisited by famine again and again. We all know how a poor and swollen population is forced by tragic compulsion to degrade its environment. We know also that the greater the pressure on the environment, the quicker it is degraded.

SCOPE could be using its scientific know-how to back up some obvious managerial points; for example, the simple fact that it is in nobody's interest to allow a large number of developing countries to become poorer and poorer partners in trade. A nation that enters the cycle of debt, inflation, unemployment and recession is degraded in every way. It becomes a real burden on the world community.

I believe that SCOPE can use its reputation for objectivity to hammer home all these ideas; it is best placed to explain that there is no better investment than an investment in prosperity. You cannot promise a return on investments straight away, but your respected voice can point to the danger and futility of simply feeding one generation of famine victims without ensuring that the next generation will not starve. Research could be concentrating on the web of concerns which sustains the environment. Without a wide perception of these intermeshed concerns, famine will continue to be seen as a function of drought alone, and the issues of family planning, land-use patterns and fuelwood consumption will be lost on the wider public, as will the implications of these losses for the richer areas of our world.

Environmentalists and scientists addressing environmental issues have a duty to demonstrate and to keep demonstrating the advantages that accrue to all nations from environmentally sound sustainable development. Sustainable development is advocated in some circles, but it has not yet become the common currency of decision makers. For some of us it has become an article of faith: often used, ofted quoted, but little explained. It is in danger of becoming a textbook theory, rather than a management tool. If we expect a better dissemination of the message of sustainable development, then the public needs to understand and see implemented a number of principles, including:

1. Help for the poorest, and food for the hungry. Those who are starving must be fed, but beyond that there must be 'bottom up' development.

2. Our 'contract with nature'. For too long we have talked of conquering nature, when really our goal should be to work with nature to improve the quality of life. Self-reliant development can only occur within the constraints of natural resource systems.
3. The idea of cost-effective development using a system of accounting that can quantify the benefits of enviromentally sound development and the costs of overexploiting our natural resources base.
4. The application of technologies appropriate for the self-help of developing countries, and appropriate for the pressing issues of land use, health, pest control, clean water and shelter. It is time to consider our priorities: before we strive to conquer outer space we should be able to feed our people on earth.
5. An insistence that human beings are the resources in question. Capital formation and economic growth alone cannot be our ultimate goals without reference to the development of human resources and social welfare.

In UNEP we consider these to be the foundations of sustainable development. We look to SCOPE to help us fine tune our own ideas. I would very much like to see if SCOPE considers it worthwhile to build on these foundations, and to develop a philosophy of resource management based on 'the human connection', based on a belief that without responding to popular need and ensuring popular participation in development, monitoring, assessment and research become irrelevant.

Resource destruction is rarely stopped by technology alone. We firmly believe that people and development go together. To demonstrate the point, PR China and South Korea – two countries with wide political differences – have both made some startling progress in reforestation. In both cases, success stemmed from popular participation. Looking at the failures, you can choose any number of failed development schemes, in any number of countries, of any political persuasion: a lack of genuine local support or understanding is a common denominator.

Beyond this philosophy of people and resources comes the question of turning a philosophy into action – into sustainable development. Research and academic enquiry can sometimes filter into the popular consciousness and into decision making circles, but not quickly and not effectively. SCOPE is ideally placed to use its energies and its influence to put these ideas across to a broader audience.

UNEP has been experimenting with reaching out to sectoral interests not traditionally concerned with the environment. Last year we were pleasantly surprised by the response of such interest groups as legislators, non-governmental organizations concerned with development, and industrialists. This year we are concentrating on women, youth and religious groups. Scientists are represented in all these groups. Members of SCOPE may wish to consider how they could be influencing these groups when adopting their policies, and moulding their activities to help sustain development and encourage the rational use of natural resources.

The more we succeed in getting various groups of society to perceive the environment as part and parcel of their responsibility, the closer we get to our goal of managing our environmental resources properly and regaining much of what has already been lost. With existing knowledge and with such support, we can stop the advancing deserts and regreen many of the areas already lost; we can replant trees and use the existing forests more wisely; we can combat marine pollution and restock depleted fisheries; we can stop soil erosion and make better use of the land already being farmed; we can prevent air and water pollution and improve the quality of life in mushrooming cities. Above all, we can tackle poverty – a root cause of many of our environmental ills.

However, all this cannot be achieved without a much more practical orientation and a better understanding of what the environment should mean to the public manager and decision maker. When I say practical orientation, I am talking about development strategies that mean something at the grass roots level. We have begun, for example, to question whether technology must always be employed in the service of making things bigger and faster, or whether it should necessarily be very sophisticated. SCOPE can help in the effort to assist governments, especially in developing countries, to take development out of the abstruse arena of hi-tech and bring it into the village and onto the farm.

We have developed many technologies that are far less harmful to our environment than they were 20, even 10 years ago. We are also able now to calculate the long-range consequences of technology. What lies within our grasp today is the power to use our mastery of science and technology to shape a better future. But we still display insufficient will and little confidence in exercising that controlling power. And we have failed dismally in the task of setting worthwhile social goals for the use of modern technologies. We have had a tendency to surrender too often to technological imperatives, instead of striving for desirable human values. It is no longer a question of sitting back and seeing where technology is taking us; it is now a matter of using science and technology to take us where we want to go. It is time to affirm that we are in charge. Complex and daunting as the problems may seem, they are of our making and are accessible to our solutions.

Technology should be a tool for sustainable development and an aid to resource management. Instead it is often divorced from the realities of development. There is a tendency to hide behind scientific uncertainty when the public calls for stricter environmental controls. More than a few governments, for instance, have been dragging their feet on the regulation of the emissions that give rise to ozone modification and to acid rain. SCOPE has the reputation for serious research and weight of evidence, and for standing by its results and reaffirming its research findings to be able to say, 'no, we do not know exactly how SO_2 and which NO_x emissions are destroying our forest and fresh water ecosystems, but we do know enough to insist on particular actions to address specific problems'. This of course calls for prudence and careful judgement, but these have become synonymous with the name of SCOPE itself.

Certainly, we urgently need to fill blanks in our understanding of the environment, to clear up the environmental imponderables, such as the likely consequences of modifying the ozone layer, and of the build-up in the atmosphere of carbon dioxide and other trace gases. Likewise, we are scarcely better informed about the marine environment. These are examples of gaps in our knowledge base which we need to fill for any proper environmental or resource management.

At UNEP our most recent effort to get technology working to secure better information for resource management has been the installation of the Global Resources Information Database (GRID). Using breakthroughs in geographical information system technology, satellite data analysis and computer hardware and software, GRID will aggregate information on population, agriculture, weather and a wide range of natural resource variables for each specific geographical location. With the improvements in data quality, supported by solid scientific backing and with an ability to feed back information to country planners in a usable form, we can help both developed and developing countries in following world-wide and local environmental and natural resources trends. This should improve and speed up development and environment-related decision making. We feel this is an area that SCOPE could be moving into.

Already SCOPE's research has provided it with much of the background for new work on management and technology. I can think of four areas, at least, where SCOPE initiatives could offer real promise to those of us concerned with the management side of environmental research.

The first of these is the greenhouse effect. Assessment of the consequences for climate of atmospheric build-up of CO_2 and other trace gases is coming along well. But assessment is not an end in itself. At UNEP we believe that the field requires more imagination, with a broader look at the social and economic consequences of a projected global warming, and a closer investigation of the possible moves to regulate emissions that contribute to the greenhouse effect and to respond to changes that could be induced by a global warming.

These are open fields, awaiting a breath of fresh air from an organization that is willing to ask questions about the implications of a global warming. My own view is that two main aspects remain almost entirely unexplored. The first is the cost of adapting to climatic change. The scientific community has so far been willing to put forward the possibility of drastic changes in the world's climate, but seems unwilling to commit itself on how seriously these possibilities should be taken, or on what action they required. If we are to accept that a medium-term warming is inevitable, what should we be doing to prepare for it? What areas need more research? What does the public need to know? The second aspect starts from the opposite point of view. What effect on the projected global warming would the regulation of CO_2 and other trace gas emissions have? If we began regulating such emissions in the near future, would that be sufficient to slow down the start of or mitigate the projected greenhouse effect?

UNEP's recent paper 'on the practical implications of the carbon dioxide question' is almost the only voice in what should be an energetic debate. I feel that some of SCOPE's work could be channelled into the missing aspects of the greenhouse debate. Your recent publication on the environmental consequences of nuclear war is a clear testimony to the authority, thoroughness and objectivity that had always characterized your work. It is definitely a marked addition to the large body of publications that have come out in so many countries over the past several months. It is significant in that it is the collective effort of top-level scientists from all over the globe. A similar effort in the area of the socioeconomic impacts of a global warming induced by the greenhouse gases would certainly help in answering some of the questions I have just raised and fill in some of the many gaps in our knowledge in this field.

This is also the type of approach that could be useful in tackling the acid deposition dilemma. Until now, scientists have not been willing enough to tell governments that the time has come for a drastic reduction in the emissions and flux of both sulphur dioxide and oxides of nitrogen. When governments have retreated behind the smoke screen of scientific uncertainty, the scientific community has too often been willing to go along with them.

Time and again, SCOPE has shown its ability to pass sombre and impartial judgement on scientific questions. We need to see that judgement applied to wider managerial issues. Two months ago, at a ministerial meeting in Helsinki to sign a protocol on the Reduction of Sulphur Emission, I asked delegates to consider the problem of acid rain from the point of view of no action. I pointed out that we now know the costs of uncontrolled sulphur emissions. We can count that cost in terms of forests destroyed and lifeless water. I questioned why some governments are reluctant to take positive steps to improve the situation when the cost of doing nothing is so high.

In all of the muddle of scientific data, a few simple facts shone through in that protocol. First, the combustion of fossil fuels gives rise to SO_2 and NO_x emissions that can be readily transported through the atmosphere across national frontiers. Secondly, there is an accepted link between these pollutants and the tragic and widespread destruction of natural resources, such as forests, soils and waters. Thirdly, the only known way to stop the destruction of these resources is through limiting the emission of the polluting agents. Of these facts there is no doubt, and it is these points that formed the basis of the protocol.

Those of us who received our training in the sciences often find ourselves bogged down in the detail of what we do not know, rather than concentrating on doing something about what we do know. The Helsinki meeting, in spite of scientific uncertainties, has taken a first concrete step to head off the destruction: a protocol to reduce by 30% over the next eight years national sulphur emissions or their transboundary fluxes. Presumably, a similar protocol on nitrogen emissions and fluxes will be forthcoming. SCOPE can help identify other areas for possible similar action.

The third issue I can see you fruitfully addressing is the plunder of the world's tropical forests. This is a perverse phenomenon. In Latin America, for example, development agencies continue to finance extensive ranching in forest areas with entirely unsuitable soils. Organizations such as SCOPE could challenge that approach on technical grounds. Medium-term investments in the intensive utilization of more suitable land away from the jungle frontier would undoubtedly offer greater return and a more stable ecological balance. It is not just in Latin America that the rape of forests continues so needlessly. In South and South-east Asia, alternatives exist to slash-and-burn agriculture, but not enough scientific talent has been put into exploring the options.

So far we have been satisfied to assess the situation, and to note that, on the basis of our 1980 assessment, on average, a minimum of 7.5 million hectares of closed tropical forest are being completely cleared and turned to other uses every year. At this rate, an eighth of all the world's tropical forests will be gone by the end of the century. Scientists have been less keen to move beyond assessment and to propose what should be done about it.

Could you be asking the question, why destroy something that can benefit you forever? Why indeed? It is an appalling comment on the world's accounting system that a nation should be said to be recording high growth rates – that is, becoming wealthy – by destroying the very foundations of its prosperity. Resource destruction is deficit spending: balance sheets should be made to reflect that simple reality. SCOPE has the expertise to work effectively on this subject.

A similar question confronts us over the freshwater issue. A bulk of research is now telling us that groundwater recharge in parts of the Middle East and North Africa has reached critically low levels. In Europe and North America, freshwater systems are being strangled by acidic fall-out. In Asia, urban water demand is drawing on many aquifers at a wholly unsustainable rate, and in Latin America a number of water supply systems are at risk as the great forests give way to economic growth. My fear is that these assessments are going largely unnoticed, because we have not been bold enough in identifying the social, economic, political and even military impacts of these trends. An organization like SCOPE can help immensely with the proper presentation of this issue.

Earlier on in this statement the need for research to fill in the large gaps in our knowledge was stressed. Several years ago, the former President of SCOPE, Professor Gilbert White, and myself appealed to scientists to exert more concerted effort towards a better understanding of the bio-geochemical cycles. We also appealed to governments and private organizations to support such an effort. Certainly there has been some response, but we are still far from fully understanding these cycles, let alone their interactions. I wish to reiterate this appeal and to focus attention on the role the findings of such scientific endeavours can play in encouraging proper environmental management and in utilizing technology for sustainable development. I doubt there is an organization better equipped to continue stressing this point than SCOPE.

Finally, I would just like to reiterate that SCOPE was among the first bodies to conceive of the importance of environmental management. We all expect you to follow that vision through. Sustainable development needs a strong advocate with a clear methodology. In my view, the view of a scientist standing somewhere between the two worlds of science and public affairs, to thrash out that methodology and apply it to some of the areas I have pointed out could be one more of SCOPE's major achievements in its untiring effort to benefit humanity.

32

Climatic effects of carbon dioxide and other trace gases

Statement to the Joint WMO/UNEP/ICSU Assessment of the Role of Carbon Dioxide and Other Radiatively Active Constituents in Climate Variations and their Associated Impacts

Villach, Austria, October 1985

The world's climate is in a continuous state of change. Until the last two or three generations, the human impact on the climate was negligible. The agricultural and industrial revolutions have changed all that. We may already have reached the point where the activities of four and a half billion human beings constitute the main motor of change. There is no point at all in debating whether it is a good thing or a bad thing that human activity can affect the Earth's climate. As the Joint WMO/UNEP/ICSU Assessment's recent studies reaffirm, for the time being we have to accept that this is a fact of life.

To date, human intervention in the world's climate has been almost entirely inadvertent. Over the coming decades that picture may change. As our understanding of the greenhouse effect grows, it may well be possible to control the industrial emission of greenhouse gases and thereby slow down the anticipated global warming. But, for the moment, we do not know for sure that such a course of action could be implemented or, if it was implemented, whether it could be effective.

We have to start thinking a great deal more seriously about the possible impacts of current trends. As a matter of urgency we must regularly review monitoring and research developments. If we fail to do this, we run the risk of being overtaken by events, and of having to deal with a global warming, for better or for worse, when it is already too late to do anything about it or deal with its impacts.

You will be familiar with past developments regarding this problem. It has been known since the turn of the century that a build-up of carbon dioxide in the atmosphere could, in theory, affect the world's climate. It has been the role of scientists to measure and predict that effect. By the late 1960s it was confirmed that the concentration of CO_2 in the atmophere was indeed rising, and that human activities were largely responsible. By using increasing amounts of fossil fuels and by burning off the world's forests, we were releasing trapped carbon into the atmosphere as carbon dioxide. In this situation a first step was to improve monitoring and assessment capabilities.

The monitoring and assessment stage is well underway. As early as in the Study of Man's Impact on the Climate (SMIC) report of 1971, some preliminary model calculations were ready. It was estimated that by the turn of the century the annual global mean temperature would rise by about 0.5°C. It was predicted that about the year 2030 the concentration of CO_2 in the atmosphere would have doubled, and that the global annual mean temperature would have increased by 2°C or thereabouts.

The main concern at that time was to begin a programme that would give us a clearer picture of the path ahead. The Stockholm Conference on the Human Environment reflected that concern in 1972 when it called for increased monitoring and research into carbon dioxide build-up. In the years following the Stockholm Conference, the United Nations Environment Programme, WMO and ICSU combined to place the study of the 'greenhouse effect' on a more sound scientific footing.

A major problem facing a concrete assessment of the phenomenon is the uncertainty over global energy demand and projected fossil fuel requirements in the coming decades. The oil crisis of the 1970s, for example, dramatically affected many existing climate-change models. Scenarios for future energy demand had to be adjusted downwards. But later scientific findings changed the picture much more dramatically. In the wake of reduced global energy demand in the 1970s the first CO_2 conference in Villach in 1980 estimated that in 2025 the atmospheric CO_2 concentration would be about 450 ppm. This model extrapolated that the time required for CO_2 concentration in the atmosphere to double would be almost 100 years – a very different time frame from the one outlined in 1971 at the high noon of the oil bonanza. Further research brought us a full circle. The discovery of the role of other trace gases has brought many scientists back to a position not far from the one spelled out in the SMIC report with respect to the projected date for doubling of the greenhouse effect.

I am aware that the debate continues as to the effect of industrial emission regulation on the rate of climatic change. I am also aware that scientific consensus at the present seems to favour a more cautious view. I know that some experts argue that even drastic cuts in fossil fuel use would delay a projected global warming by only a limited number of years.

With well directed research we can find out what effect specific emissions regulations would have on the climate of coming decades. As a precursor to entering that difficult arena, UNEP, WMO and ICSU agreed in 1982 on a stepwise investigation of the CO_2 problem. Having done this we have again gathered in Villach to consider the results of an extensive assessment effort that has been carried out by the International Meteorological Institute in Stockholm with the support of UNEP, WMO and others.

A number of salient points seem to emerge from these studies. First, the concentration of CO_2 in the atmosphere continues to increase. Net emissions of CO_2 from biota due to deforestation and land-use change are not expected, in themselves, to cause a significant change in climate, although they contribute to the rate of increase.

Second, some other trace gases in the atmosphere, in particular methane, nitrous oxide and other oxides of nitrogen and chlorofluorocarbons, have similar greenhouse effects on climate as CO_2. As they also have been found to be increasing in concentration due to human activity, their warming effect must be added to the one caused by CO_2, and we must examine the option of cutting back on these emissions as necessary.

Third, earlier studies which considered CO_2 alone suggested that levels of CO_2 concentration would have doubled from the preindustrial period by some time after 2050. It is now estimated that, by adding in the warming effect of other trace gases, the equivalent of such a doubling may occur as early as 2030. Trace gases seem to be playing a much larger role in bringing about a greenhouse effect than was earlier expected.

Fourth, we have now laid aside most of the doubts as to the effect of the build-up of CO_2 and other trace gases on global climate. The IMI assessment has confirmed that there is almost unanimous agreement that the global average surface temperature would increase in response to a doubling of the greenhouse effect. Differences in the amount of increase arrived at through the application of various modelling approaches are modest, in fact insignificant for our current purposes. It is now clear that scientists are reasonably confident that, at current rates of build-up, a global mean annual temperature increase of several degrees will probably occur over the next half century.

Following on from this are the implications of such a change for the Earth's ecosystems. We cannot yet predict with any great accuracy regional patterns of climatic change, but there are indications that there may well be a radical redistribution of the world's productive croplands. Such an impact could also alter the pattern and frequency of El Niño events and drought, as well as the rate of desertification. Its net effect may even be beneficial in some cases. At this stage, we just do not know. However, such changes could have an immense, but as yet incalculable, impact on the global political system.

Of equal or even greater significance will be the effect on the world's oceans. The predicted thermal expansion of the ocean water body alone could lead to a rise of the world's sea level of about one metre. The Arctic ice cap could disappear at the summit and the west Antarctic icesheet break away from its continental mooring. Nearly a third of all human beings live within 60 kilometres of the coast – thus hundreds of millions of city-dwellers from Bangkok to New York – would be affected, as would the port facilities that sustain the world's trade.

The picture is still clouded with uncertainty, but broadly we can say that a climatic change of the predicted magnitude would have enormous social and economic consequences that defy the imagination.

What is required from a scientific community keeping in close touch with the economic and political realities is an agenda for action. The greenhouse phenomenon is not simply an issue for the North. Its scope is certainly global, and there is increasing evidence to show that a number of developing countries are likely to be major contributors to the expected climatic warming, especially in the early decades of the next century, and many developing countries will certainly be affected. So, in the preparation of the agenda for action, both developed and developing countries are expected to participate actively.

Four elements to be considered in this agenda stand out:

1. We must examine in more detail the options being placed in front of the world's decision makers. There is a need for a serious discussion between governments and industry on the feasibility of reducing industrial carbon dioxide and trace gas emissions. There is certainly a need for a wider debate on such issues as the costs and benefits of a radical shift away from fossil fuel consumption. There is also a need for an evaluation of the political and economic costs of not taking such action. It is important to carry out a study of which industries are involved in the emission of greenhouse gases, and to begin an investigation of possible actions to be taken by particular industries on a cost–benefit basis. Most importantly, we should start in earnest a serious debate on what sorts of socioeconomic impacts we should be prepared to live with and develop clear options for the decision makers to choose from in responding to the potential impacts of climatic change.

2. We must have a commitment to further scientific and technical research. Climatic models and other projections must be improved greatly if they are to be a credible foundation for political action.

3. What will the greenhouse effect mean for the person on the street 20 or 40 years from now? To date, there has been almost no attempt to provide the public and its representatives with any real understanding of the social and economic consequences of the anticipated warming. At the moment we could only provide the barest of ideas. But it would be a start – a foundation to build upon and a context for increasingly informed debate.

4. Is there a need to create a machinery to set this ball rolling? I invite discussion on the possibility of establishing an international coordinating committee on greenhouse gases. Such a committee could encourage and review monitoring and research developments in such areas as those I have just mentioned, and also issue statements to governments, international organizations and the public at large regarding the need for particular actions and the options open in response to a potential warming of the global climate.

The greenhouse problem brings science to a new dawn. In spite of a host of uncertainties, it remains clear that the world's climate is now subject to the intervention of human beings. Debate must focus on how best to handle this intervention. By this, I mean challenging available resources in a way that will allow us to understand, anticipate and possibly even direct changes in the global climate for the benefit of humanity. If we have the power to alter the climate, why should we not harness that power for the good of humankind? It is an exciting prospect. Scientists can lead this debate.

33

Desertification control – moving beyond the laboratory

Statement to the Arid Lands Today and Tomorrow Conference

University of Arizona, Tucson, Arizona, USA, October 1985

Desertification is a human tragedy. Rooted in human mismanagement, it awaits a human solution. Its advance is neither unpredicted nor unstoppable. And yet the deserts are on the move. It is a tragedy that locks the greater part of one billion people into a cycle of poverty and destruction. It is a catalyst for famine and for conflict. It feeds upon itself, and it promises still greater destruction in the years to come. Its causes, however, are largely unredressed; its effects are misunderstood; and the tools to bring about its end lie around us, largely unused and sometimes unnoticed.

We all know that no lands are more vulnerable to desertification than the arid lands. No soils are more fragile and more in need of good management than those of the drylands. Mismanagement or no management at all are the basic causes of desertification: they underlie the famine in Africa, and explain the loss of billions of dollars of agricultural production. It is on this topic – the relationship between desertification and management – that I wish to address my remarks. My message to this conference is a suggestion that our efforts to halve desertification might be partly misdirected. So much of our effort goes into technical areas, and I wonder if this is wholly justified or satisfactory.

When I draw up a mental balance sheet, I become puzzled. On the one hand I see the mass of assembled talent here today. Few disciplines are as well served by men and women of such high scientific calibre as the multisectoral study of arid lands. Represented at this conference we have scientists from every continent, and of every political persuasion. We have here a global cadre of excellence mustering to address a global threat.

Looking now at the other side, things are not so bright. You are all aware that as much as 6 million hectares of agricultural land are estimated to be lost to desertification every year. An additional 21 million hectares are reduced to zero economic productivity. There is no evidence to suggest that next year will be any better; it seems that, even as more talent is brought to bear, the problem worsens.

When thinking over this discrepancy one feels that we may have let the study of arid lands become too much of an academic pursuit. Having dealt with the problem of desertification for the past 10 years and having been involved in one way or another, for several years before, in the consideration of the problems of arid lands in general, I have come to believe that an organized sortie into the future handling of arid lands issues and of countering desertification will mean a move from the confines of the laboratory and the textbook into a world of politics, economics and social forces. It is a move long overdue, and I applaud those of you here who are showing us a lead.

The world over, there are scores of able technicians looking into arid lands and desertification. Yet we are not achieving any meaningful progress. We appear to be tackling these problems without enough emphasis on the basic requirement of mobilizing the motor forces of public concern and political will. The 1977 United Nations Conference on Desertification is a case in point. It is generally recognized, in technical and scientific terms, to have been one of the UN's best prepared conferences. It brought together many of the world's leading authorities on desertification and elicited from them a Plan of Action to Combat Desertification. The Plan, laid down the specific targets that would have to be reached in various areas to arrest the problem. It spelled out in some detail the steps we would have to take to reach those targets within a set time frame. Technically, international experts agreed it was a feasible plan, soundly based and ready for implementation. At the decision-making levels, governments applauded the Plan and readily endorsed it.

Almost a decade down the line, where are we with this Plan? 'Nowhere' is perhaps an exaggeration, but it is the word that comes to mind. UNEP's frustrations are symptomatic of the marked lack of success the experts have had in tackling desertification. Of course, there are successes we can point to, some of them quite important. But they represent a few drops in the bucket, no more.

Something is going wrong with our efforts. We have the technical know-how to beat back the deserts; we have a time frame and a costing for action and for *inaction*. We estimate that the cost of stopping desertification world-wide would be in the region of US$4.5 billion annually over a period of 20 years. Already governments and international and other organizations spend an estimated US$2.1 billion a year in this effort. But of the US$2.4 billion needed additionally every year only US$0.6 is available. The deficit is US$1.8 billion per year. For five years in a row we have failed to meet any part of this deficit. Compare this, on the other hand, to the close to US$1000 billion spent every year on arms. Or compare it more closely to the realm of our work, to what we are losing yearly in agricultural products – the equivalent of US$26 billion.

Why then, is so little being achieved? Supported by able scientists and engineers, proven technologies and sympathetic governments, we should be well on the way to halting desertification. The answer, I believe, lies with our failure to convince the public. There is, you might say, no perception of threat. In some ways the term 'drought' has deflected the public's attention from the real danger.

Political decisions are not made in a void. Decision makers – and politicians most of all – formulate and adjust their plans largely in response to public opinion. Without the weight of public pressure, issues – however important – tend to slip out of sight. People do not care because they do not see how the problem can affect them, nor do they see a way out of the web of circumstances within which they live. Governments have, therefore, little or no mandate, and sometimes no ability or willingness to act.

I am not going to proselytize with you here. You are among the few who do see the problem. I would rather share with you your present perception of the desertification issue. It is not a purely technical angle: it is an attempt to see desertification in the wider context of environmentally sound management. It is an attempt to give a context for public concern and participation and for government action.

To us in UNEP, sustainable development seems to be the most viable context for coordinating popular appeal and directing management action to combat desertification. In broad terms we see the concept of sustainable development encompassing:

1.   help for the poorest because they are left with no option other than to destroy their environment;
2.   the idea of self-reliant development – within natural resource constraints;
3.   the idea of cost-effective development using different economic criteria to the traditional approach; development should not degrade environmental quality, nor should it reduce productivity in the long run;
4.   the great issues of technologies for good health, appropriate food, education, clean water, and shelter for all;
5.   the notion that people-centred initiatives are essential; human beings, in other words, are *the* resources in the concept.

A consensus on the means to achieve sustainable development is also emerging and can be summarized as follows:

- raising indigenous environmental and natural resource management capability;
- building upon experience in the North and learning from their mistakes of the past;
- ensuring that environmental considerations are not left out in development planning;
- gathering sufficient hard data of an environmental kind (taxonomic, ecological, geological) to enable sound development planning to occur;
- concentrating on systems particularly at risk, be it arid lands, watersheds, moist forest or areas of rapid urban expansion;
- most important of all, informing the public of what is at stake.

Each point is linked with the others. Plans for countering desertification or managing arid lands in general will have to confront squarely the problems of overpopulation, illiteracy, bad land-use patterns and energy, just as surely as they are going to involve sand dune fixation, improved irrigation and genetically improved crops.

To set this process of sustainable development in motion we must convince the public and their leaders. These leaders are not, for the most part, people with a background in the physical or natural sciences. Decision makers tend to be economists, lawyers, social scientists, bankers, businessmen, planners and the like. Because of their immediate responsibilities these people do not make decisions with the interests of the next generation in mind. Their decisions are made with a view to the next election, the annual balance of payments or to the next meeting of shareholders. We have to convince these decision makers that an investment in better arid land management, in desertification control and in other forms of environmental protection pays. To do this, we must emphasize two crucial interrelationships.

The first is the critical relationship between the natural world, which includes human society, and its development. Conservation of natural resources and social development must be pursued together as goals of equal importance. Interactions between social development forces change: resources, environment, people and development make it necessary for governments to think in terms of trade-offs between alternative courses of action.

The second interrelationship is the connection between development and economic growth. Recent years have witnessed a broadening of our understanding of this relationship. Development is no longer seen exclusively as a matter of the growth rate of national income or the rate of capital formation. The new emphasis is on wider and more qualitative aspects of development, such as improvements in income distribution, employment, health, housing and education.

Applying these principles to the areas of arid lands and desertification control, we find ourselves having to address problems that are not of an immediately scientific nature, but which have enormous impact on the affected ecosystems. The issue of land tenure, for instance, goes to the root of our problem. We believe that for irrigated and rain-fed agricultural land, farmers must have outright ownership at the individual or collective level before they will devote the energy and the resources needed to conserve the land.

Poor land management should be in the focus of our attention. There must be realistic incentives for sustainable land use. In the developed countries, tax incentives and tax penalties can be used to bring economic interests into line with environmental ones. In the developing world these methods are less easily applied. Here, governments could reward environmentally sound land-use practices, such as terracing, and building check dams, with price subsidies for agricultural implements and supplies. Depending on the situation on the ground, authorities could be providing seed, fertilizer, agricultural implements, water pumps and so on, free, or at discount prices as a reward for applying conservation measures.

Other incentives could include awarding development projects preferentially to areas which follow accepted guidelines on good land use, improving extension services, credit schemes and working within agricultural cooperatives to promote the benefits of sustainable development.

Another option would be 'production mode regulation'. Sensible penalties can be brought to bear on those who unnecessarily degrade their land. In the developing world this could take the form of farmers who use bad land management being the last to receive support from agricultural credit schemes, both from central authorities and from the level of the cooperative and local administration.

These are just a few ideas: ideas that I know many of you have been discussing for some time. They are ideas that might cast some light on our lack of success in countering desertification. Desertification control is a multisectoral issue involving a range of scientific disciplines, technical and engineering skills, as well as social and economic management. It is not that we have been concentrating too much on the purely scientific issues, it is that we have not concentrated enough on the social, economic and political aspects of desertification control.

As a closing remark I would call upon my colleagues participating in this conference to use the upcoming workshops to discuss some of the social and economic aspects of their work. Technical work in every area carries with it a host of opportunities, and a baffling array of constraints. This is where we could be concentrating some of our energy.

To touch on just one example, say, rangeland management, we have, on top of technical solutions, the need for education and training, adequate family planning, the need for improved tenure systems, options for alternative means of production and many others. Lined up against these we see resource constraints, political, language and cultural constraints, economic constraints, and, of course, time constraints. Once we begin working on these as really integrated packages, I think we will be making real headway.

The state of our arid lands and the advance of desertification is almost entirely out of public consciousness. As long as that continues the problem will probably worsen. Recommendations of able scientists, qualified planners and skilled engineers will remain unimplemented for as long as it takes for the public to come on board.

We will never get them on board until we dig out the real causes for their remaining aloof when they see destruction of the natural resource base on which their own lives depend so clearly written in front of their eyes. It is only then that the highly qualified technicians will be able to devise applicable means of arid land management and of desertification control, based on a clear understanding of the real constraints at the local, national, regional and global levels.

So in these workshops, can we begin by breathing life into an issue that has remained, in part, closeted in the halls of academe? Can we seriously consider why the affected public is not acting to head off the destruction? Can we explore in depth the social and economic options and constraints that follow from antidesertification schemes? Can we investigate thoroughly the political and popular dimensions? Can we try and carry some perception of the issues to statesmen, decision makers, and the informed public? Can we tie the arid lands question into their own concerns?

There is a mass of talent here working on arid land problems. A fair number of these men and women are moving out into the web of concerns that surrounds its technical centre. It is a web that embraces local populations, decision makers and the general public. With this pool of excellence, I see no reason why we cannot make more headway in identifying the whole spectrum of constraints facing desertification control and charting the options open to overcome them. The challenge is to do this at the local level with the many local variables. I am certain your discussions of these wider issues will yield the much needed guidance to our collective effort in this field.

34

Towards an African Solution

Opening Statement to the First African Ministerial Conference
on the Environment

Cairo, Egypt, December 1985

A generation after independence Africa is still not free – not free from poverty and illiteracy, and not yet completely free from the economic chains that have bound her since colonial times. But at least there are signs that African solutions are emerging to the African crisis. This Conference is one of them. Africa is a bountiful continent, blessed with a wealth of resources. But the continent's natural resources – its grasslands, forests, croplands, freshwater and fisheries – are being squandered.

Never before has there been such an assembly of African ministers responsible for the environment. The convening of this assembly is a clear recognition, not only that many African nations are confronting environmental bankruptcy, but also that those nations are ready to face it together.

Throughout the continent there are examples which give hope and inspiration – programmes and projects which show that destruction can be prevented. We are not starting from scratch. The Organization of African Unity, the Economic Commission for Africa and many UN agencies have worked closely with African governments to develop practical regional action programmes in their fields of competence. Across the continent highly qualified insitutions and experts are at work; projects are being implemented; a number of African Commissions and networks are functioning and further plans are being actively formulated for concrete subregional cooperation. Nor is the report before you the brainchild of UNEP and its partners OAU and ECA. It is the digest of two years of consultation among African experts, between UNEP and the experts, and between UNEP and our sister agencies in the UN system.

The Conference Report and several back-up documents provide a comprehensive assessment of the nature and extent of the environmental crisis confronting the peoples of Africa. We can be in no doubt that the renewable resources of this continent are being dismembered at a truly alarming rate. But there is insufficient awareness within Africa and beyond of just how alarmed we should be. Consider the following:

- The forests are shrinking by four million hectares per year. In some countries in West and Central Africa the rate is slowing because there is virtually no tree cover left.
- Many species of wild animals and plants are threatened and some are on the verge of extinction.
- Overuse and misuse of Africa's ancient and fragile soils are causing soil erosion on a catastrophic scale. Over 60% of the land area north of the equator is susceptible to erosion.

- The last 100 years have seen a 150 km-wide belt of productive land on the southern edge of the Sahara Zone turn completely unproductive. Since 1968, one quarter of Africa's semi-arid pasturelands – the main source of meat – has also been rendered unproductive.
- Africa's rich fishing grounds are being overfished and coastal regions are threatened by pollution.
- Africa's river systems are coming under increasing pressure from pollution, watershed destruction and overuse. Now only two areas, the Guinean Zone and the equatorial zone, have a net surplus of water.

The roots of Africa's enduring crisis lie in the interaction of a number of complex factors: rapid population growth, an unfavourable international economic situation, harsh climatic conditions, and in many cases unsustainable development policies.

Over the past few years the pattern of destruction has unfolded in a broadly similar way across Africa. In other continents intensive agriculture has allowed food production to keep pace with population growth. In Africa there has been a wholesale neglect of rain-fed agriculture and a failure to make adequate investment in inputs such as improved seeds, fertilizers and irrigation. To feed a rapidly increasing population the only alternative has been to expand the area under cultivation and to practise extensive rather than intensive farming.

We know the results. Semi-arid land which should be sparsely grazed is being overgrazed; steep slopes and other marginal lands are going under the plough; vegetation cover needed to protect drainage systems is being cleared; whole communities are being displaced.

Africa has the world's lowest life expectancy, the highest rural underemployment and the severest food shortages. These are the social and economic consequences of Africa's environmental crisis, but they are rarely seen as such.

As people are caught up in the cycle of destruction, more and more Africans are forced to compete for dwindling resources. Ultimately, the security of nations, both internal and external, is threatened. The reaction of some governments has been to invest more in arms. Of 37 African countries for which data are available, only 10 spend more on agriculture than on the military. This does not achieve meaningful security.

Africa's derisory share of the world's industrial output has left her appallingly vulnerable to a punitive international economic system. Low commodity prices, the debt burden, markets pre-empted by protectionism and unfulfilled promises have put a break on development. Too seldom do we consider how these factors can result in environmental deterioration.

Africa is also suffering from the impact of apartheid. In South Africa, the Banthustan policy has starved black farmers of investment, forcing them to degrade the land – their own food base. This is one more incentive to rid the continent of the last vestige of racism.

If Africa is the only continent in the world where per capita food production has failed to keep pace with population growth, governments cannot apportion all of the blame to outside forces. In a new cooperative endeavour, African nations must find viable solutions. This is what we are here to do. Before this Conference convened, a group of leading African experts met to lay the groundwork. They agreed that there already exists a wide consensus on what needs to be done.

In the Lagos Plan of Action, African nations have a strategy which could be made to work if only the political will existed. As the OAU's Council of Ministers pointed out, if the Plan had been implemented, 'the ravaging effects of the current world recession and drought on African economies would certainly have been minimized'. Africa has many of the networks, agreements, and institutions required to implement the Lagos Plan.

The expert group considered that our goal should be self-sufficiency in food and energy. There is reason for confidence that this can be achieved. Almost half of all the land suitable for stock rearing or cultivation is underutilized. Just 10% of all the fresh water returning to the seas around Africa would be enough to irrigate 13 million hectares, thereby alleviating hunger.

When it comes to energy, Africa has scarcely begun to tap its vast potential. A fraction of the continent's hydroelectric potential is being used, and geothermal scarcely at all. Fuelwood accounts for over 80% of Africa's energy use, but most is squandered on open hearth fires. Meanwhile, accessible technologies – improved cooking stoves, solar, biogas and wind and animal powers – which could eliminate the drudgery of firewood collection, have scarcely moved beyond the experimental stage.

Another problem is the lack of a coordinated policy among the many hundreds of aid and relief organizations, each pursuing in good faith, and with enthusiasm, their own blue-prints for recovery. There is a danger that the best intentions could further aggravate problems.

Home-grown solutions, which combine tradition with the latest know-how, have shown that the wastage of the resource base can be stopped. The solutions are not applied across the continent because the mechanisms to spread them either do not exist, or more often are underutilized.

Africa therefore has the potential for recovery. Self-sufficiency in food and energy can be achieved. In reaching for this goal I would particularly stress the need to:

1. safeguard traditional systems such as nomadism and fallow practices which respect the limitations of Africa's sensitive environment and adapt these systems through new techniques to achieve the increased production levels needed to support higher levels of population;
2. give special attention to the very poor and provide them with know-how, low-cost inputs and economic incentives which will help them conserve their resources;
3. promote with greater vigour national family planning programmes;
4. say no to development assistance which benefits the relatively well-off at the expense of the most disadvantaged;
5. do much more to pool resources and information on solutions that are in place and working.

To ensure success, the rural majority, the villagers themselves, must be involved. Experience has shown that time after time even well designed projects break down because they fail to take account of the social and cultural realities. Conservation works only if the villagers can readily appreciate the benefits.

To apply this, the report before you suggests and your experts recommend the following action. First, they recommend the establishment or activation of continent-wide networks of institutions. The networks should cover the major elements necessary for environmental rehabilitation – monitoring, climate, energy, soil, water, genetic resources, science and technology and education and training. They would pool the available African wealth of talent, experience and know-how and each would serve Africa in its entirety.

Second, a serious effort should be made to implement a number of subregional programmes. Most of them have been very thoroughly planned and components of a number of programmes are in fact ongoing. But there is a need for fresh resolve. Delays, inefficiencies and problems created by political differences must be overcome.

Third, it is recommended that two regional pilot projects should be mounted: one covering 150 villages and the other 30 semi-arid stock-raising areas. The target would be to help these communities become self-sufficient in food and energy. The cost would be reasonable. Such concrete projects would help us to target better international support. The lessons learned would be invaluable. Self-sufficiency in these project sites could be achieved within a reasonable time, probably five years, but this will depend on careful project design involving national government departments, aid agencies, relevant UN organs and organizations and, of course, the people concerned themselves.

I am aware that the world is watching the outcome of this Conference. I am confident you will take concrete, realistic and implementable decisions. To guarantee the implementation of these decisions your experts recommend:

- that this Conference of Ministers be institutionalized, meeting every two years to review action taken and decide on new action;
- that the Conference establishes four technical committees on deserts and arid lands, forests and woodlands, rivers and lake basins, and seas, to work between sessions to oversee the implementation of the Conference's decisions;
- that the Conference authorizes its President and Bureau to take the necessary action between sessions to ensure proper follow-up of Conference decisions.

In cooperation with the ECA and the OAU, UNEP is ready and willing to provide a Conference Secretariat.

African governments will be expected to bear the costs of their participation in the follow-up to this conference. A budget of some US$11 million per year, in convertible currencies, will be required. Governments will be further expected to share their experience through in-kind contributions to the network operations and to the pilot and demonstration projects. Governments may wish to call upon the UNDP to finance their foreign currency contributions to a Trust Fund to be established for this purpose, from the development assistance presently made available to them. This will show the world your real commitment to implementing what you decide upon. It will ensure effective participation by all in an African cooperative effort. I am confident that the proposed regional and subregional activities will constitute a focus for meaningful international cooperation. We are encouraged by the wide representation of bilateral aid agencies, non-African governments and the UN system. The rehabilitation of Africa's economy is high on their agenda and they certainly have a crucial role to play in helping to implement the decisions of this Conference. But in the final analysis, the decision is exclusively yours – and the main burden of making the decisions work will fall to African countries.

The world community has responded with great generosity to famine relief in Africa. It is your responsibility to offer concrete proposals for cooperation in the medium and long term. Africa can recover, through its own local efforts and with the genuine and well oriented support of the outside world. We have in Africa the means and the will to make the high expectations of African nations at the time of independence once again a reality.

The proposed programme has been worked out by your own experts – it is an African solution to an African problem. Five hundred million Africans need programmes that will place solutions to the African crisis firmly in African hands. I am sure you will accept your responsibility and rise to the challenge.

35

Socioeconomic benefits of environmental protection

Keynote Address to the Belgian Royal Academy

Brussels, Belgium, June 1986

In this address I intend to depart from the norm. I want to begin my statement by citing two examples which I believe will set the agenda for my chosen theme, the social and economic benefits of environmental protection.

The Minnesota and Michigan Mining Company, better known as 3M, is a highly profitable US transnational company. A strange choice, you might think, to begin making the case for the socioeconomic benefits to be gained from protecting the environment. But what 3M has shown during the last decade in the USA is that there are significant profits to be made from reducing and recycling wastes, instead of treating and discharging them. By reformulating and redesigning processes, 3M eliminated each year 90 000 tonnes of air pollutants, 10 000 tonnes of water pollutants and 150 000 tonnes of solid wastes, saving into the bargain about US$200 million. The economic benefits are axiomatic, and the social benefits, in terms of a cleaner human environment, scarcely less so.

The second example comes from the Himalayas where watershed management is bringing cost–benefit ratios over the medium term (up to 10 years) of 2.5 : 1. In Nepal, incomes of farmers participating in agro-forestry schemes have risen four-fold over a similar time span.

These examples from the industrialized and developing worlds raise the curtain on the theme I should like to consider, namely that the future health of the economy and the well-being of the people who depend upon it are inextricably bound up with the fate of the environment. The central issue is not whether to choose between development and industrialization and preserving the environment, it is how to select the right ways to develop, ways that will not only minimize damage to the world about us, but that are also actually designed to enhance the quality and productivity of the environment, and hence the economy.

What has begun to filter through to decision makers and the public is that we cannot support, let alone improve, living standards by depleting the productive base, meaning human and natural resources. Without more investment in health care and education and without economic development that sustains soil cover, forests, fisheries, atmospheric quality and energy resources, the future is discounted.

There is also a growing realization that there is no damage limitation to environmental deterioration; in some cases it is having a global impact. Take the issue of global warming. Between 1950 and 1980 the CO_2 pumped into the atmosphere by factories, power stations, cars and the burning of the forests, increased by 0.4% per year. This, with other trace gases, is inducing the global warming. In Austria last year, a conference of the world's leading scientists in this field agreed that in little over a generation the world will experience a temperature rise of anywhere between 1.5 and 4.5°C representing the temperature change of the previous 125 000 years. It is a source of deep concern to UNEP that so little concerted action is planned to head-off what could turn out to be a catastrophe – one aspect of which would be increasing sea levels, thereby inundating coastal areas where a third of the world's population live. The pattern of world agriculture could be altered, with serious consequences for the political stability of our planet.

Though the future impact of nations' economic activities on the climate may be difficult to gauge, this is manifestly not the case with the encroaching deserts. Desertification is an ugly word for an ugly phenomenon. The size of the areas at risk is 35% of the world's land surface. Between a quarter and a half of this is already seriously affected. Every year 6 million hectares are reduced to a state of complete uselessness and 21 million hectares – more than seven times the size of Belgium – deteriorate to the point where they are no longer economically productive. About 850 million people – 20% of the world's population – live in the areas at risk.

The lost production has been valued, in 1980 prices, at US$26 billion per year. Nine years ago, meeting in the aftermath of the terrible Sahelian drought of 1968–74, the world's nations agreed a plan to stop the desert's march. The cost, again in 1980 prices, would be US$4.5 billion per year, a benefit–cost ratio of more than five to one. Yet, close to the 10th anniversary of the Plan of Action, a special UN account set up to finance antidesertification measures has received the derisory sum of just over US$200 000. In our 1984 survey we found that virtually nowhere has the situation improved.

I have mentioned climate and desertification to illustrate the global nature of the threat to the biological foundations of the world economy. I could equally have chosen tropical rainforest destruction – within a single generation between one third and one half of the existing tropical rainforest cover could be lost; or species extinction – over the same period we could witness a holocaust of living things, depriving the continents of 10% of their species for ever; or the threat from acid rain – particularly in North and Central Europe, North America and now gaining a sinister foothold in Turkey, India and other developing countries; or the deteriorating living conditions in the towns and cities of the Third World – a social and political time-bomb.

All these disturbing trends are closely linked. We do not yet properly know how the cogs of the environment machine, the terrestrial ecosystems, work. We know much less about the two thirds of the planet that is water. What we can say is that some teeth in the cogs of our environmental machine have been destroyed and that increasing numbers are sustaining damage. At what point will so much damage be done that the machine will break down? We simply do not know. But we do have some warning signals. The most dramatic, and the most tragic, has been the recent famine in Sudan and Ethiopia. Drought triggered the famine but its underlying cause was ecological destruction. It is estimated that the famine-stricken Highlands of Ethiopia are losing soil at the rate of 2000 tonnes per square kilometre each year. This is the crux of the problem which no amount of food aid parachuted or trucked into famine-stricken areas can overcome.

There is widespread agreement that something must be done. As the two examples cited at the start of my address reveal, action is being taken, but despite the obvious social and economic benefits, these continue to be the exception rather than the rule. Why? The problem organizations like UNEP address develop over decades. They are often unseen, with remote rural populations or the teeming millions in the shanty towns at the cutting edge of the global averages. Unless riots or a famine break out, their effects are difficult to bring home and are not immediately felt. If a plant species, which might have properties of immense pharmaceutical or industrial value, becomes extinct another tooth in one of the many cogs is lost, but no one notices because the machine continues to function.

No standard cost–benefit analysis can measure this kind of loss. There is no convenient guide for negotiators and decision makers on what they can do and how they should react. Tackling the problem of environmental deterioration by taking preventive and ameliorative action often involves foregoing today's tangible economic benefits for tomorrow's intangibles. But tackled they must be, difficult as they may seem. Many require urgent action, even though the benefits will only be felt in the long term.

The payoffs from a switch to better management of resources are quantifiable. The present costs of switching to better policies and programmes are small in comparison to future costs or the damages that will result if insufficient action is taken. The industrial world is now finding that the costs of recycling poisonous wastes are insignificant in comparison to the clean-up costs of inadequately regulated hazardous waste sites. The US government, for example, is in the process of agreeing a US$10 billion budget to render these dumping sites safe.

The developing world is now finding that the costs of watershed protection are also small in comparison to the costs of increased flooding and reduced irrigation and hydroelectric capacity. The urgent need is for states to reckon the value of intact resources in their national accounting systems. For example, it is estimated that each year Brazil's Amazon rainforest produces goods – rubber, fish, brazil nuts and so on – valued at close to US$100 million. Assessing its value as rain-maker (the Amazon generates half its own rainfall through evapotranspiration), as genetic storehouse and as freshwater supplier is extremely difficult, but it may amount to billions of dollars per year.

The message is clear – overuse of resources must be replaced by conservation. The oil price hikes of the early 1970s promoted energy conservation. Today the average manufactured product in industrialized countries is made with 20% less energy than a decade ago. As the examples of FR Germany and Japan show, economic growth can be sustained even if energy consumption drops. We need to get into the habit of including the environmental dimension in the main indicators of wealth creation, such as food, technology, energy, commodity prices, financial transactions and borrowing. Take the first item on my list – food. Agriculture is much more than an economic activity, it is also a social and environmental activity. Investment in irrigation will increase productivity, but without environmental controls, for how long? Inundating land with water will increase production, but in a tropical climate particularly, if environmental safeguards are not built in, the evaporating water will suck up salts, leaving an arid hardpan. It has been estimated that in the Sahel, as fast as new land goes under irrigation, existing land goes out of production.

This need not happen. Through such modern methods as spray and drip irrigation, using available renewable energy sources, these problems can be avoided. In Iraq and Oman, major programmes are underway to revive traditional irrigation systems, which at one point seemed doomed to extinction due to the diesel pump. These countries are finding that traditional methods, combined with the latest know-how, can be more productive – and socially desirable – in the long term.

The point is that for virtually every environmental problem I have outlined, there is a technological and scientific fix available, and the promise of something better tomorrow. So where is the blockage? I return to my earlier observation that promoting environmentally-sound development involves foregoing today's tangible economic benefits for tomorrow's intangibles. The task at hand for us in UNEP, and in the other multilateral organizations, is to make those intangibles tangible; to bring the future into today's calculations of profit and loss.

So what is meant by the term 'intangible'? Consider the lack of sanitation and safe drinking water – the world's most acute pollution problem, though it is seldom seen as such. According to the World Health Organization, lack of clean water and decent sanitation is responsible for 80% of all world disease. Waterborne diseases kill about 25 000 people every day and debilitate millions more every year. This is a severe brake on development. A survey carried out in Venezuela proved this dramatically, in reverse. When clean water was provided in the countryside, production increased so much that the cost of installing the water supplies was recouped five to seven times over – intangible into tangible.

Much the same can be said for insect-borne diseases, in particular malaria to which nearly half the world's population is prone. In Kenya's coastal zone, at any one time, one in four people is suffering from a bout of the disease. In Africa South of the Sahara, a million infants are killed each year by malaria. According to one estimate, the cost of developing and marketing the elusive vaccine would probably amount to less than a dollar per life. The economic returns in terms of productivity and social well-being would be immense. Again this is changing intangible into tangible.

The bottleneck is that it is the poor who would benefit immediately from the improvement in health. The rich nations and the rich elite of the developing nations would benefit in the long term. Therefore, although the Brandt Report and other studies, too numerous to mention, make an unimpeachable case for economic self-interest, the tangible at the level of policy implementation remains intangible.

It was said of the French radicals in the 1920s and 1930s that although their 'hearts were on the left their wallets were on the right'. A similar observation can be made of most governments today, for although the rich nations and the priviledged elite in Third World countries subscribe to equity, their economic and political conduct remain little changed. Indeed, there are disturbing signs of a move away from development assistance and of a widening gap between wallet and heart. The effect has been to grind the poor. This starts at the most exalted levels of international finance, working its way through the system. Developing nations, for example, have had to export increasingly larger amounts of natural resources and cash crops to import the same amount of products, or to service their external debts. One Latin American country, for example, had to export nearly 10 times as much beef in 1981 to buy a barrel of oil as it did in 1973.

The consequence of a failure to reform international economic relations to tackle global inequity has been to deepen poverty, and poverty has long been recongized as the number one agent of environmental destruction. Poverty degrades the human environment and, in doing so, obstructs development. Poor villagers will continue to fell the trees, overgraze and overcultivate the land and have the largest possible family when they see no hope of a better future. It is a lifestyle of self-immolation.

At the other end of the scale, the industrialist and politician will go for short-term gain when he sees only the next balance sheet or the next election. This too, though in a far less obvious way, is self-immolation.

Yet we have those two examples – 3M and the Nepalese farmers. These give us hope that the intangible can become a factor in economic management. The opportunities for improving resource management, environmental quality and standards of living, at both ends of the economic spectrum, are enormous.

At the larger 'bottom' end, we need more low-cost, simple programmes that address the neglected needs of neglected populations. There are numerous examples to show that low-cost health and family planning programmes, emphasizing basic services and preventive care, can reduce mortality rates and raise contraceptive use by a factor of two within five years.

In Mauritania, an inexpensive NGO agro-forestry scheme, placing a premium on community involvement, boosted millet production by 60% over seven years. Close to Nairobi, a women's tree-planting project has poor mothers enthusiastically involved in tree planting and protection. Too few perhaps, but these are success stories nonetheless. The common denominator in each has been community participation, stimulated not by some benign desire to improve their environment, but by a need recognized by local communities to gain control over local resources.

The revolution in information technologies has presented us with the opportunity to spread the message and to show the billions of poor what they can achieve for themselves through environmental protection. Here we have another excuse for optimism.

It is changing the policies of those who decide and implement them that must draw our fire. I am talking not only of those in government, but of the industrialists, financiers, religious leaders, educators, planners and economists. These are the people whose decisions can turn the intangible into the tangible, who can put the heart and the wallet in the same place. They can create a political and economic climate in which projects and business ventures, large and small, can be designed for the sustainable benefit of the majority. It is a challenge to the scientific community and to academia to refine the means of showing the real benefits to the decision makers. It is a hard job, but it has to be done.

My conclusion, therefore, is that, when decisions take account of the environment from the outset, social and economic benefits will accrue in the medium to long term. To do this we need the tools. This is the responsibility of those who can provide them – social, economic, natural and physical scientists.

This is a question of ethics, but much more so of economic and social self-seeking. The English philosopher, John Locke, put it well when he stated that no appropriations of natural resources were valid unless they 'leave as much and as good for others'. We have the example of 3M and the Nepalese agro-forestry programme to show that we can utilize resources without imperilling our own welfare and that of our children.

Abbreviations

ASEAN	Association of South-East Asian Nations
CEC	Commission of the European Communities
CIDIE	Committee of International Development Institutions for the Environment
CILSS	Permanent Interstate Committee for Drought Control in the Sahel
CITES	Convention on International Trade in Endangered Species
ECA	Economic Commission for Africa
ECE	Economic Commission for Europe
ECLA	Economic Commission for Latin America
ECOSOC	Economic and Social Council of the United Nations
ECWA	Economic Commission for Western Asia
EEC	European Economic Community
EMEP	European Monitoring and Evaluation Programme
EPA	Environmental Protection Agency of the United States of America
ESCAP	Economic and Social Commission for Asia and the Pacific
FAO	Food and Agriculture Organization of the United Nations
GEMS	Global Environmental Monitoring System
GRID	Global Resources Information Database
IAEA	International Atomic Energy Agency
ICC	International Chamber of Commerce
ICSU	International Council of Scientific Unions
IFAD	International Fund for Agricultural Development
IFIAS	International Federation of Institutes for Advanced Study
IIASA	International Institute for Applied Systems Analysis
ILO	International Labour Organization
IMI	International Meteorological Institute
INFOTERRA	International referral system for sources of environmental information
IPPF	International Planned Parenthood Federation

IRPTC	International Register of Potentially Toxic Chemicals
IUCN	International Union for the Conservation of Nature and Natural Resources
IWC	International Whaling Commission
IYSH	International Year of Shelter for the Homeless
NIDS	New International Development Strategy
NGO	Non-Governmental Organization
OAS	Organization of American States
OAU	Organization of African Unity
OECD	Organisation for Economic Cooperation and Development
SCOPE	Scientific Committee on Problems of the Environment
SMIC	Study of Man's Impact on the Climate
SWMTEP	System wide Medium Term Environment Programme
UNCHS	United Nations Centre for Human Settlements (Habitat)
UNCOD	United Nations Conference on Desertification
UNCTAD	United Nations Conference on Trade and Development
UNDP	United Nations Development Programme
UNEP	United Nations Environment Programme
UNESCO	United Nations Educational, Scientific and Cultural Organization
UNFPA	United Nations Fund for Population Activities
UNIDO	United Nations Industrial Development Organization
USAID	United States Agency for International Development
WCS	World Conservation Strategy
WFC	World Food Council
WFP	World Food Programme
WHO	World Health Organization
WICEM	World Industry Conference on Environmental Management
WMO	World Meteorological Organization
WWF	World Wildlife Fund

Index